Mapping the Skies

Mapping the Skies

ROBIN KERROD

BARRON'S

Library of Congress Catalog Card No. 00-102324

ISBN 978-0-7641-1587-5
ISBN 0-7641-1587-1

All inquiries should be addressed to:
Barron's Educational Series, Inc.
250 Wireless Boulevard
Hauppauge, New York 11788
http://www.barronseduc.com

This book was designed and produced by
Quintet Publishing Limited
6 Blundell Street
London N7 9BH

Creative Director: Richard Dewing
Art Director: Sharanjit Dhol
Design: Ian Hunt
Senior Project Editor: Toria Leitch
Map Illustrations: Danny McBride
Illustrations: Richard Burgess

Typeset in Great Britain by
Central Southern Typesetters, Eastbourne

Manufactured in Hong Kong by
Regent Publishing Services Limited
Printed in China by Winner Printing and Packaging Ltd

19 18 17 16 15 14 13 12 11 10

CONTENTS

Introduction

These days astronomers can pinpoint particular stars or other heavenly bodies in any part of the heavens with ease and guide their large and powerful telescopes to them with infinite accuracy. Precision instruments, computers, and astronomy satellites have all helped make mapping the heavens a precise art.

Yet the basic concept behind producing star maps dates back thousands of years to the dawn of civilization. It is the concept of the celestial sphere, the great dark dome that appears to surround the Earth, on the inside of which the stars are fixed.

Astronomers pinpoint stars on this imaginary sphere—in the heavens—by a system of celestial coordinates akin to the latitude and longitude system used to pinpoint places on Earth. This provides the basis for astronomers' star maps.

The night sky does not stay the same all the time. Constellations come and go as Earth circles in its yearly orbit around the Sun. Our maps chart the march of the constellations by showing the changing views of the sky month by month throughout the year.

We use a three-stage approach to exploring the night sky each month. First, sketch maps help you get your bearings—there are sets for both Northern and Southern Hemisphere observers. Second, the monthly star map shows the major constellations in some detail. And third, the accompanying text provides background information and highlights features of particular interest.

The accompanying planisphere is a further aid to familiarizing yourself with the heavens. It can display the stars visible at any time of the night and on any date of the year.

When you are out in the field allow some time for your eyes to get accustomed to the dark first. And then use the little red flashlight provided to read the maps and planisphere. The red light won't affect your night vision like an ordinary light.

Then gaze at the heavens to your heart's content; enjoy the greatest free show on Earth. And, as astronomers everywhere will echo—may you have clear skies.

Robin Kerrod

Robin Kerrod on location, at Kitt Peak National Observatory, Arizona ▼

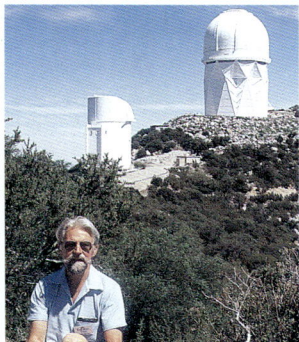

Spotting the Constellations

Although there is no actual celestial sphere, astronomers assume there is one when they come to map the heavens and locate the stars. They pinpoint a star on the sphere by a similar system to that used in geography to pinpoint a place on the surface of Earth—latitude and longitude.

In geography, the latitude of a place is a measure of the distance that place is north or south of the Equator. Its longitude is a measure of how far around the world it is from a fixed point. The latitude and longitude are both measured in degrees. Latitude is measured in degrees from the Equator, longitude from a north–south line, or meridian, running through Greenwich in England. It is called the Greenwich meridian.

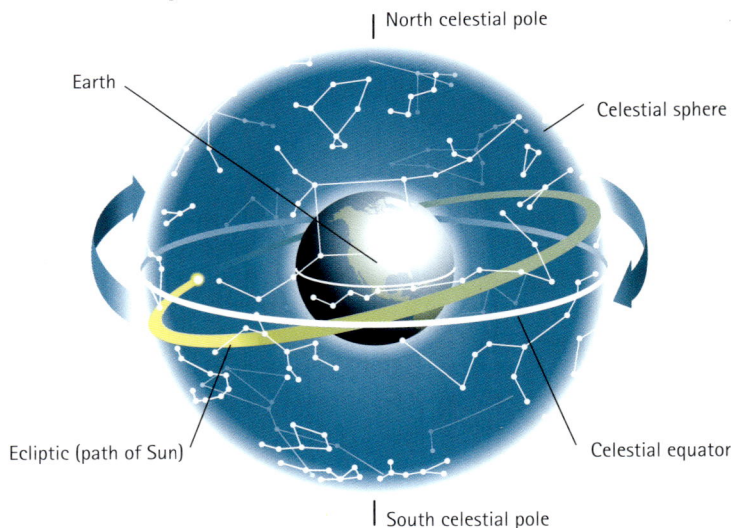

North celestial pole

Earth

Celestial sphere

Ecliptic (path of Sun)

Celestial equator

South celestial pole

CELESTIAL COORDINATES

The system of celestial latitude and longitude works on the same principles. Celestial latitude is a measure of how far a star is north or south of the celestial equator. The celestial equator is a circle around the middle of the celestial sphere, just as the Equator is a circle around the middle, or "waist," of Earth. Celestial latitude is called declination (symbol δ Greek delta). It, too, is measured in degrees. Degrees north of the equator are deemed positive, while those south of the equator are negative.

CELESTIAL LONGITUDE

Celestial longitude, like its terrestrial counterpart, is a measure of the distance around the celestial sphere a star is from a fixed point. In this case the fixed point is the place where the Sun appears to travel across the celestial equator in the spring. Or, in other words, it is the point where the ecliptic (path of the Sun) and the celestial equator intersect. This sounds more complicated than it is. The diagram will help make it clearer.

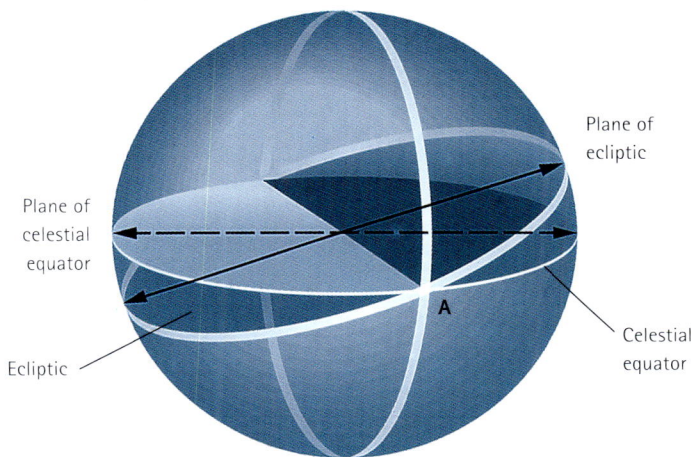

Plane of ecliptic

Plane of celestial equator

A

Ecliptic

Celestial equator

The point of intersection is known as the First Point of Aries. The celestial longitude of a star is measured from this Point around the celestial equator to a great circle passing through the star and the celestial poles. Again, this sounds complicated, but the diagram will clarify it.

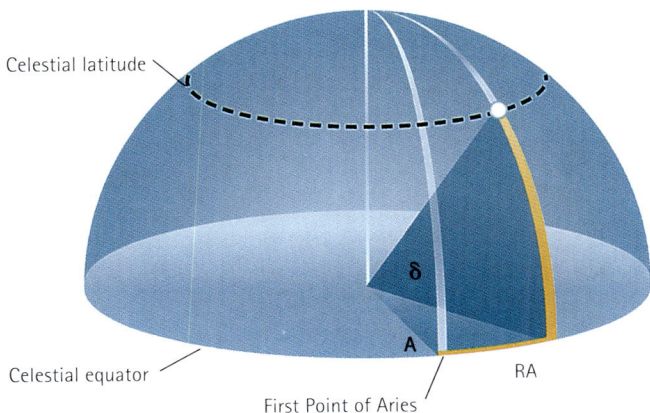

Celestial latitude

δ

A

Celestial equator

First Point of Aries

RA

Celestial longitude is called right ascension (RA). To complicate matters further, it is not measured in degrees, but in hours. And these are not the everyday hours that we set our clocks by, but hours of astronomical or sidereal time—time relative to the stars! The subject of normal and sidereal time is discussed further in the companion Star Guide.

The Monthly Maps

The stars move overhead during the night because Earth spins around in space. They rise over the horizon in the east, culminate (climb highest) at the meridian, and eventually set beneath the horizon in the west. As the months go by, different constellations appear in these whirling skies, while others disappear. This happens because of Earth's other motion in space, its orbit around the Sun.

Every month, as Earth travels a bit farther in its orbit, we look out at night onto a slightly different part of the celestial sphere. After a year, Earth comes full circle as it completes its orbit around the Sun, and we look out at the same sky that we saw 12 months before.

In this book we follow the changing aspects of the night sky literally month by month, from January through December. We show how the skies look at the same time of night and at the same time of the month, so we can see at a glance how they change.

THE SKETCH MAPS

For each monthly section, we present sets of sketch maps to familiarize stargazers with the sky that month—to help them, actually, to find their celestial bearings. They feature the most prominent constellations on display.

There are four sketch maps—two for observers in the Northern Hemisphere and two for observers in the Southern. One of each pair of

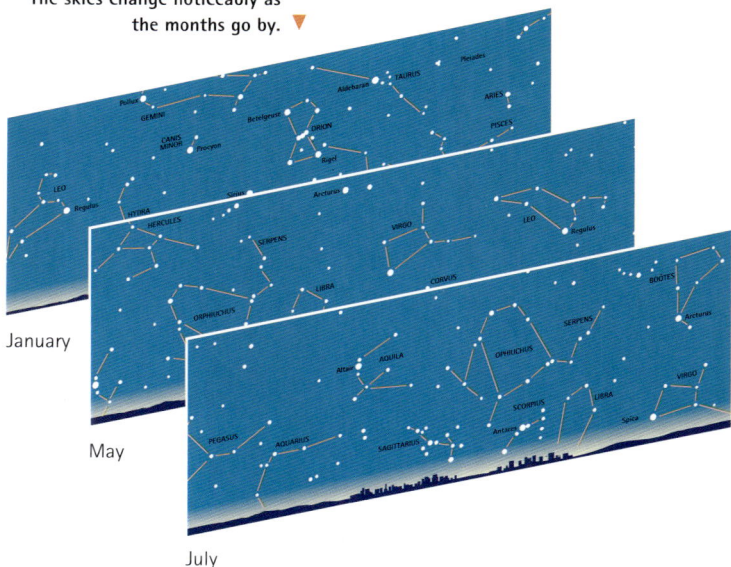

The skies change noticeably as the months go by. ▼

January

May

July

sketches shows the skies looking north, the other the skies looking south. If connected together, they would present a 360-degree panorama of the heavens that month.

THE STAR MAPS

We obtain the main star maps for the 12 months of the year by splitting up the celestial sphere into 12 segments, as shown in the illustration. The segments each form the center part of each star map. On either side are parts of the previous and the following segments. (The stars, of course, are on the inside of the segments of the "exploded" sphere, not the outside.)

The "exploded" celestial sphere. ▼

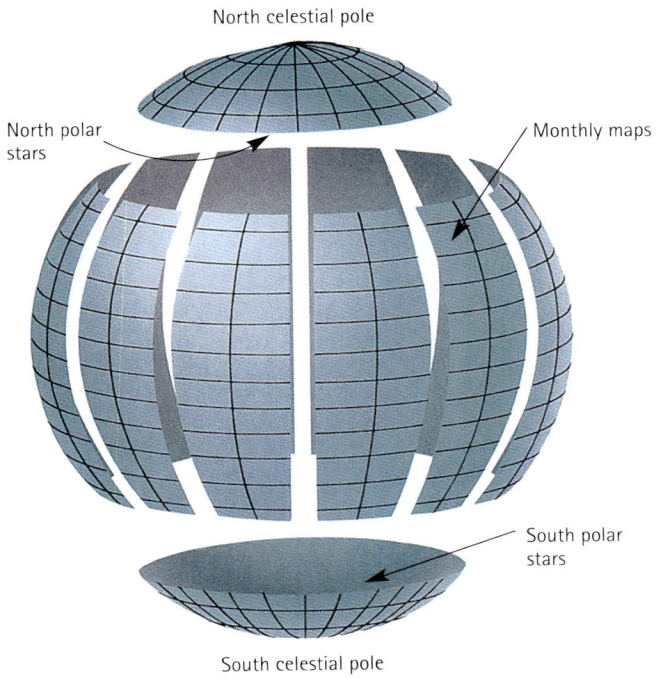

North celestial pole

North polar stars

Monthly maps

South polar stars

South celestial pole

The coverage of the stars on the celestial sphere is completed by two polar maps, one in the north and one in the south. They represent the "caps" of the exploded sphere. Many observers in the Northern Hemisphere can see some of the north polar stars all the time. We say these stars are circumpolar (see page 14). Likewise, some of the south polar stars are circumpolar for many Southern Hemisphere observers (see page 18).

The Star Maps: AN ESSENTIAL GUIDE

The main monthly star maps are drawn with a grid of celestial latitude and longitude, that is, degrees of declination on the vertical axis and hours of right ascension on the horizontal axis (see page 8). The maps show the constellations that appear near the meridian—the north-south line in the sky—at about 11:00 P.M. in the first week of the month. They may appear slightly to the east or west, depending on the exact date of observation.

As an aid to understanding the maps, we show opposite a smaller version of the map for July with annotations highlighting key features. The stars shown are ones that can be seen with the naked eye. The various sizes of dots representing stars give an indication of their brightness. The largest represent the brightest, first-magnitude stars. The smallest dots represent stars of about the fourth magnitude.

Stars may be identified in a number of ways. Some of the most prominent have proper names, such as Vega, which is the brightest star in Lyra. Others are usually identified by a letter of the Greek alphabet, the brightest in a constellation being Alpha (α), the next brightest Beta (β), and so on.

Certain symbols are used to identify what are called deep-sky objects, which include nebulae, open and globular clusters, and galaxies. Such objects are identified in two main ways, either by an M number or an NGC number. M stands for Messier and the number refers to the object's position in a famous catalog drawn up by the French astronomer Charles Messier. NGC stands for *New General Catalogue*, a listing of nebulae and clusters compiled by the Danish astronomer Johan Dreyer and published in 1888.

Map of July skies showing typical features. ▶

The Greek Alphabet

α	Alpha	ι	Iota	ρ	Rho
β	Beta	κ	Kappa	σ	Sigma
γ	Gamma	λ	Lambda	τ	Tau
δ	Delta	μ	Mu	υ	Upsilon
ϵ	Epsilon	ν	Nu	ϕ	Phi
ζ	Zeta	ξ	Xi	χ	Chi
η	Eta	o	Omicron	ψ	Psi
θ	Theta	π	Pi	ω	Omega

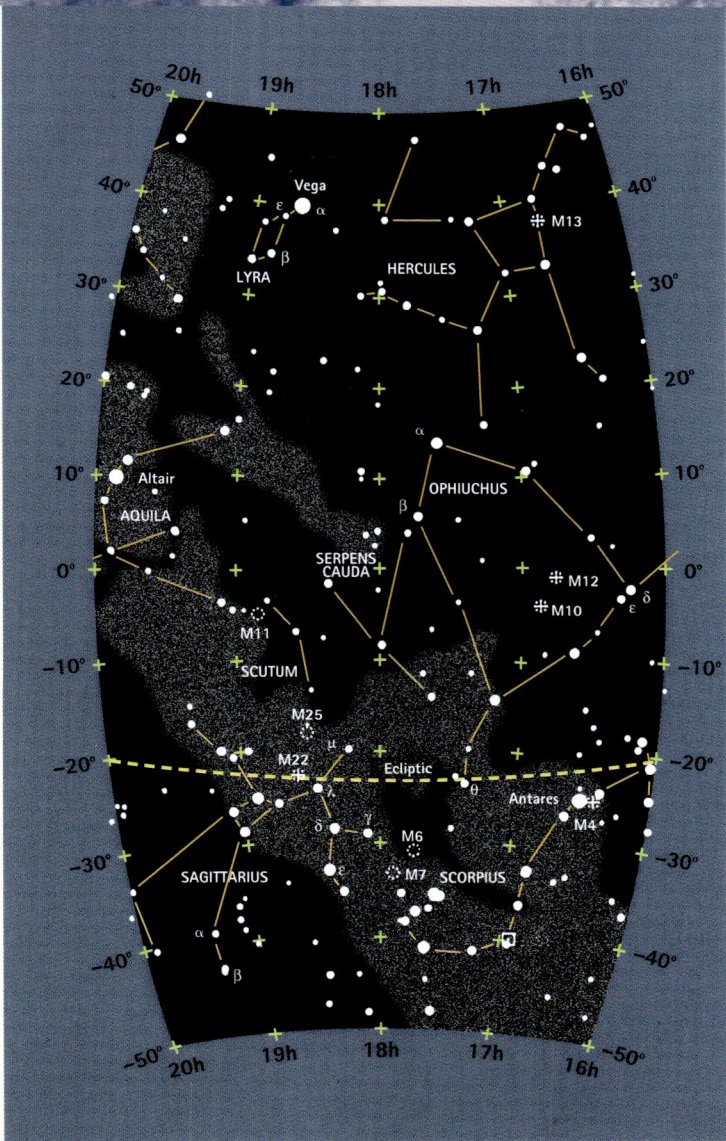

20h 19h 18h 17h 16h
50°   50°
40°   40°
30°   30°

Vega
ε   α
β
LYRA
HERCULES   ⊞ M13

20°   20°
10°   10°

α
Altair   OPHIUCHUS
AQUILA   β
SERPENS   ⊞ M12
CAUDA   ⊞ M10
0°   ε δ 0°
M11
SCUTUM
−10°   −10°
M25
μ
−20°   M22   Ecliptic   −20°
⊞   θ
δ   γ   Antares
M6
SAGITTARIUS ε   M7 SCORPIUS M4
−30°   −30°
α
β
−40°   −40°

−50°   −50°
20h 19h 18h 17h 16h

MAP SYMBOLS KEY

Symbol		Symbol	
●	Star	▢	Nebular
•	Size indicates level of brightness	⬭	Galaxy
⚬	Open cluster	▒	Milky Way
⊞	Globular cluster		

North Polar Stars

Far northern skies lack the brilliance of far southern skies, but they do possess something that the southern ones lack—a convenient Pole Star. Astronomers call this star Polaris; another name for it is North Star. This star is located nearly directly over Earth's North Pole, in line with Earth's axis of spin, and for this reason hardly changes its position at all.

CASSIOPEIA

Cassiopeia is one of the easiest constellations to recognize because of its distinctive W-shape. But more than a little imagination is needed to picture this constellation, as the ancient Greeks did, as a "lady in a chair." The lady in question was Queen Cassiopeia, wife of King Cepheus and by all accounts incredibly vain. It was her vanity that led to their daughter Andromeda being offered as a sacrifice to the sea god Poseidon.

Cassiopeia has a number of double stars, including Alpha (α) and Eta (η), while Iota (ι) is a triple star, visible in small telescopes. But, being in the Milky Way, the constellation is most notable for its dazzling array of clusters. Some are easily picked up in binoculars, such as NGC457 and NGC654, which are found south of the W. Small telescopes will pick up many more.

CEPHEUS

This is not a particularly bright constellation but can easily be found because it lies next to the distinctive W-shape of the brighter Cassiopeia. In mythology, too, Cepheus and Cassiopeia were close, being king and queen of ancient Ethiopia. Andromeda was their daughter.

Among the stars, Beta (β) shows up as a double star in small telescopes. Delta (δ) is also double; the main star is a yellow supergiant, which varies in brightness as regular as clockwork, between magnitudes 3.5 and 4.4 in precisely five days, nine hours. It is the prototype for the variable stars known as the Cepheids. An interesting thing about a Cepheid is that its period of variation is directly related to its absolute magnitude, or true brightness. So when we see one, we can work out how far away it is.

Mu (μ) also changes in brightness, but not with the same precision. It varies between the third and fifth magnitudes over a period of about two years; it is one of the Mira-type variables. But it is the color of Mu that rivets the observer. It has a lovely red hue, noticeable to the naked eye and striking in binoculars or small telescopes. It is called the Garnet Star.

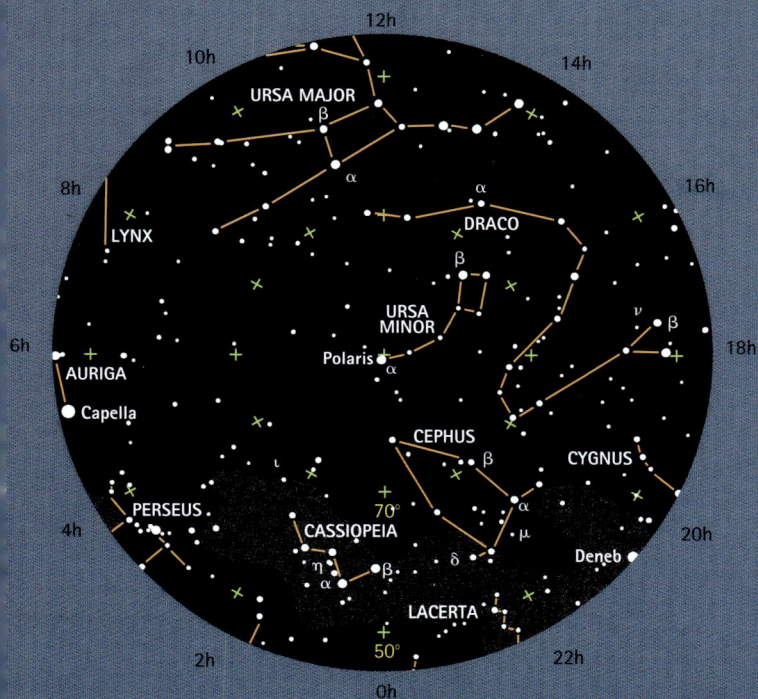

DRACO, THE DRAGON

Draco is a sprawling constellation of faint stars that winds nearly halfway around the Pole Star. It is identified with the dragon in Greek mythology that guarded the golden apples in the garden of the Hesperides, which Hercules had to collect as one of his labors. He managed to kill the dragon and pick the apples.

Of interest among the stars is Nu (ν) in the Dragon's mouth, a wide double for binoculars and small telescopes. Alpha (α), also called Thuban, is notable because in ancient Egyptian times it was the Pole Star.

URSA MAJOR, THE GREAT BEAR

This group features as a Key Constellation, see page 16.

URSA MINOR, THE LITTLE BEAR

Ursa Minor is also known as the Little Dipper because it looks rather like a ladle used for drinking. In mythology, the Little Bear represents a nymph who nursed Zeus as a baby. It is a small constellation, which contains the north celestial pole. The Pole Star Polaris lies within a few degrees of the pole and is best found by using the Pointers in Ursa Major.

Ursa Major, THE GREAT BEAR

This sprawling northern constellation is the third largest. It includes the prominent asterism—group of stars—that virtually everyone in the Northern Hemisphere is familiar with, astronomers or not. This group of seven stars is known as the Big Dipper in North America, because it resembles a ladle used to dip into a bucket of milk or water. It is called the Big Dipper to distinguish it from the similarly shaped, but smaller and fainter, Little Dipper, the constellation Ursa Minor, or Little Bear.

In Europe this star group is seen as a plow, taking the form of the handle and share blade of an old-fashioned horse-drawn plow. The asterism also has a third name, Wain (meaning plow) or King Charles's Wain, supposedly in honor of the first Holy Roman Emperor in Europe Charlemagne (A.D. 800s).

To the Greeks, Ursa Major was Callisto, a beautiful girl who was seduced by the king of the gods, Zeus. After Callisto bore Zeus's child, his wife Hera turned her into a bear.

POINTING NORTH

The Big Dipper or Plow has been a boon to navigators for thousands of years. This is because two of its stars point to the North Star or Pole Star, which astronomers call Polaris. This star remains more or less in the

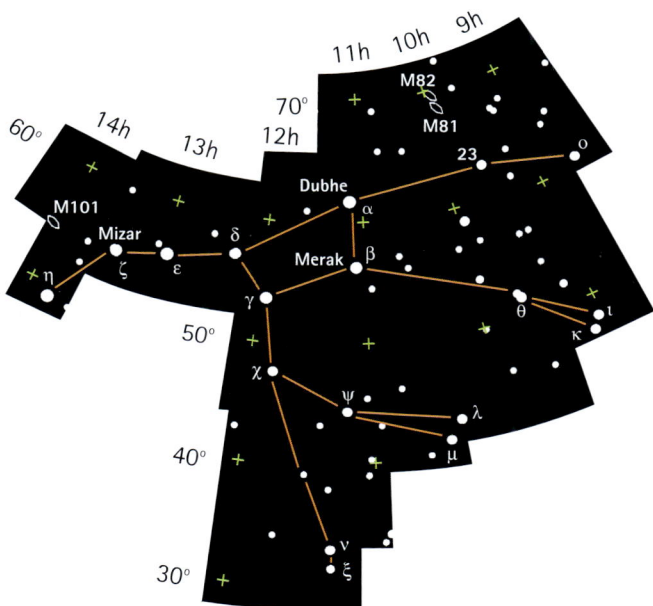

60° 14h 13h 12h 70°
11h 10h 9h

M82
M81

23

Dubhe
α

Merak β

M101
Mizar δ
η ζ ε
γ θ
50° κ ι
χ

ψ λ

40° μ

30° ν ξ

same spot in the night sky. The two Pointers are named Merak and Dubhe. A line from Merak through Dubhe will point to Polaris. The best-known other star in the Big Dipper (Plow) is the second along the handle, named Mizar. Close by is another bright star, Alcor. The two stars form one of the best-known double stars in the heavens. They make for a good eye test—if you can see them separately, your eyesight is good.

Ursa Major is located in a relatively featureless part of the heavens, which is why it stands out so prominently. But it has a few interesting objects, including a few galaxies within range of binoculars. They include M101, a spiral found by following the line of three stars in the Big Dipper or Plow handle. Northeast of Dubhe is the beautiful and bright spiral M81, a galaxy similar to our own. Seemingly entangled with it is M82.

This spiral galaxy is M106, one of many to be seen in the constellation. ▶

South Polar Stars

In the far south there is no convenient Pole Star as there is in the far north. So it is a little more difficult for navigators to get their bearings and, incidentally, for astronomers to set up their telescopes. However, the south polar region is elsewhere dazzlingly bright, containing the most brilliant region of the Milky Way and beacon stars like Alpha and Beta Centauri, Canopus, and Achernar.

CARINA, THE KEEL

This constellation was once part of the ancient constellation of Argo Navis, the ship of the Argonauts (see page 34). It lies on the edge of the Milky Way and is rich in nebulae and clusters. Its leading star Canopus is outstanding because it is the second brightest star in the heavens, after Sirius.

Among the dazzling array of stars in the constellation, Eta (η) is special. At present it is just visible to the naked eye, but only a century ago it blazed brighter than any other star in the sky except for Sirius. It is embedded in a glorious nebula, which is visible to the naked eye but best seen in binoculars or small telescopes.

The stars Iota (ι) and Epsilon (ϵ) lie close to Delta (δ) and Kappa (κ) in Vela. This quadrangle of stars forms an X-shape known as the False Cross, because it may be mistaken for the Southern Cross, Crux.

Among the many bright clusters that Carina boasts, IC2602 is the finest, located around Theta (θ). It has eight stars brighter than the sixth magnitude, and is well called the Southern Pleiades.

CENTAURUS, THE CENTAUR

This dazzling group of stars features as a Key Constellation (see page 20).

CRUX, THE SOUTHERN CROSS

The Southern Cross is the most famous southern constellation of all, even though it is the smallest in the sky. The ancients regarded it as part of the hind legs of Centaurus, and it has been recognized as a separate constellation only since the 1600s.

In the Cross itself, Gamma (γ) is reddish, contrasting with the three other stars, which are bluish white. It is a double, like Alpha (α), also named Acrux, and Beta (β). Around Kappa (κ), which is close to Beta, is one of the finest open clusters in the heavens, full of sparkling colors. Designated NGC4755, it is better known as the Jewel Box.

There are other clusters visible in this constellation, as it lies in a rich part of the Milky Way. Between Kappa and Alpha there is a dark void in

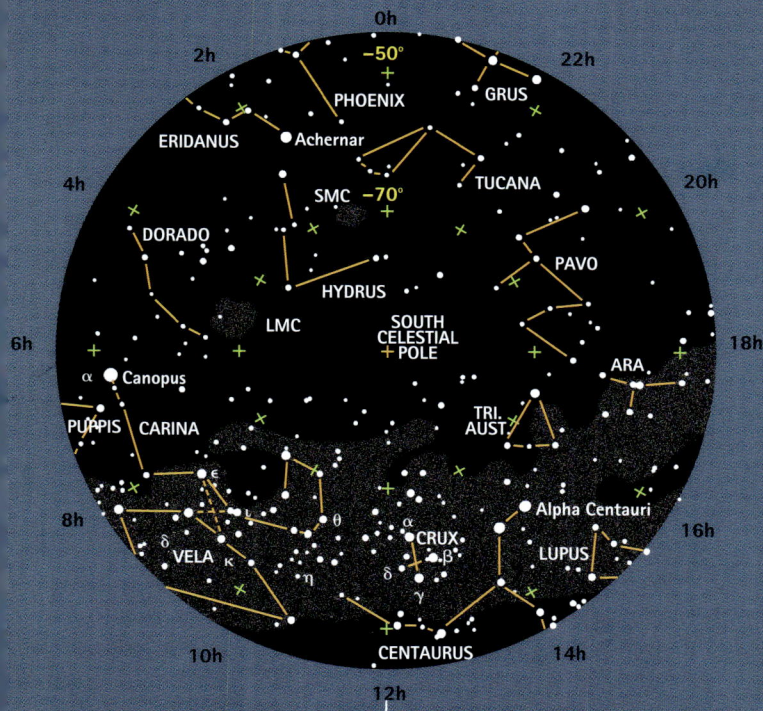

the starry vista, known as the Coal Sack. It is not a hole in the Milky Way, but a cloud of dark gas that blots the light from the stars behind.

DORADO, THE SWORDFISH

Another relatively modern (1600s) constellation, Dorado is best known for hosting the nearest galaxy to our own, the Large Magellanic Cloud (LMC). This irregular galaxy can be seen with the naked eye as a fuzzy patch at the southern end of Dorado's line of stars. It is so close to us (about 170,000 light-years) that small telescopes will begin to reveal its stars, clusters, and nebulae. The brightest nebula, the Tarantula, is even visible to the naked eye.

TUCANA, THE TOUCAN

This southern bird was also introduced in the 1600s and in turn plays host to a smaller companion of the Large Magellanic Cloud, the Small Magellanic Cloud (SMC). This too is an irregular galaxy. The Toucan's other claim to fame is the naked-eye object originally classed as star 47. But telescopes reveal this "star" to be a gigantic ball containing hundreds of thousands of stars packed together.

Centaurus, THE CENTAUR

Centaurus is one of the largest of all the constellations, and a spectacular feature of far southern skies. It surrounds the most distinctive southern constellation, Crux, the Southern Cross. The constellation is named after one of the wise centaurs in Greek mythology, named Chiron. Centaurs had the torsos and heads of men, but the backs and legs of horses.

THE SOUTHERN POINTERS

The two brightest stars in Centaurus are unmistakable pointers to Crux. The brighter of the two, Alpha (α) Centauri, is the third most brilliant star in the heavens, after Sirius and Canopus. It is also the nearest bright star to us, being some 4.3 light-years away. Small telescopes will show that Alpha Centauri is a fine binary star, made up of two components. Close by is a much fainter star, known as Proxima Centauri. This red dwarf star holds the distinction of being the closest star to us, at a distance of about 4.2 light-years.

A fainter star in the constellation, named Omega (ω) Centauri, is not a star at all, but a congregation of hundreds of thousands of stars. It is a globular cluster, one of hundreds that orbit around the center of our

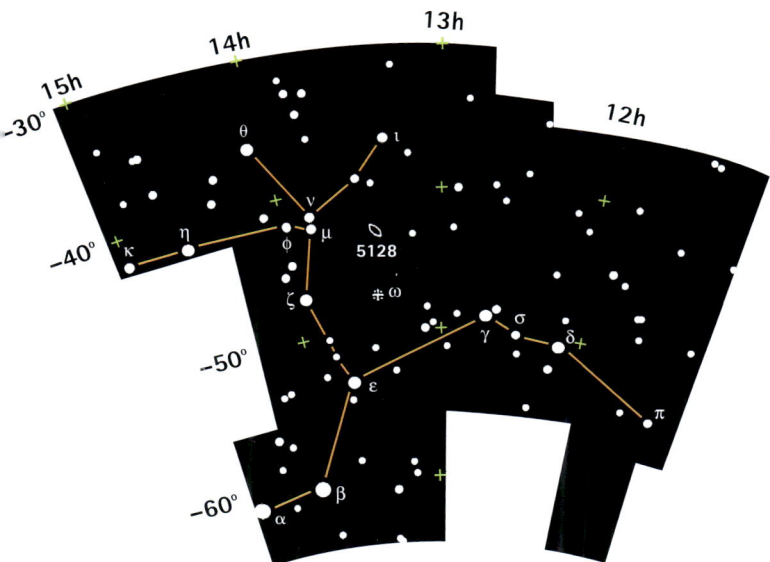

galaxy. Omega Centauri can readily be seen with the naked eye, and binoculars will begin to show what a huge object it is. Telescopes will reveal how packed together the stars are in the cluster. In the center, they are probably only about a tenth of a light-year apart.

Close to the triplet of stars in the left shoulder of the figure of the Centaur is an object easily visible in binoculars. It is not a star or a cluster, but a galaxy (NGC5128). It is not an ordinary galaxy, though. Pictures show it to be cut through by a dark dust lane. And radio astronomers have discovered that it is a powerful source of radio waves, called Centaurus A. Indeed, there are only two more powerful sources in the sky. Centaurus A puts out so much energy that it is classed as an active galaxy, and is probably powered, like other active galaxies, by a black hole.

▶ Some of the stars and billowing gas clouds near the center of the Centaurus A radio galaxy, pictured by the Hubble Space Telescope.

January Stars

January skies are probably the most stunning of the year, dominated by the majestic figure of Orion. We look at this fine constellation in detail on page 26. But its two brilliant stars, Betelgeuse and Rigel, are only two of the month's highlights. In north and south alike, brilliant stars shine down—Arcturus, Castor, Pollux, Aldebaran, Procyon, and dazzling Sirius. The Milky Way runs right across the sky, providing a feast for viewers using binoculars.

AURIGA, THE CHARIOTEER

This bright northerly constellation has the easily recognizable shape of a kite. Its brightest star is the yellowish Capella, the sixth most brilliant in the heavens. Capella represents a she-goat draped over the left shoulder of the charioteer. Just to the south are three of its kids, visible as a triangle of fainter stars.

Of the other stars, Epsilon (ϵ) is most interesting. It is an eclipsing binary, consisting of a brilliant supergiant star and a dark companion revolving around each other. Every 27 years the dark companion passes in front of the supergiant and blots out much of its light. This has the effect of making Epsilon dim noticeably. Strangely, it stays dim for months at a time. This suggests that the companion star must be surrounded by a dark cloud of gas and dust.

The Milky Way runs through the constellation, and features a number of bright star clusters, easily visible through binoculars.

CANIS MAJOR, THE GREAT DOG

This constellation may be small, but it is easy to pick out because it boasts the brightest star in the sky. This star is Sirius, also known as the Dog Star. As stars go, Sirius is not particularly bright. It appears very bright to us because it lies relatively close to us, at a distance of only about nine light-years.

Just south of Sirius is the cluster M41. On a clear night, you might even glimpse it with the naked eye as a misty patch. One star in the constellation you won't be able to spot is the so-called Companion of Sirius, also called the Pup. Sirius and the Pup circle around each other, forming a two-star, binary system. The Pup is a tiny, very dense star of the type we call a white dwarf. Historically, it is important as the first white dwarf to be discovered, in 1862.

JANUARY SKIES ▶
Constellations visible near the meridian at about
11:00 P.M. during the first week in January.

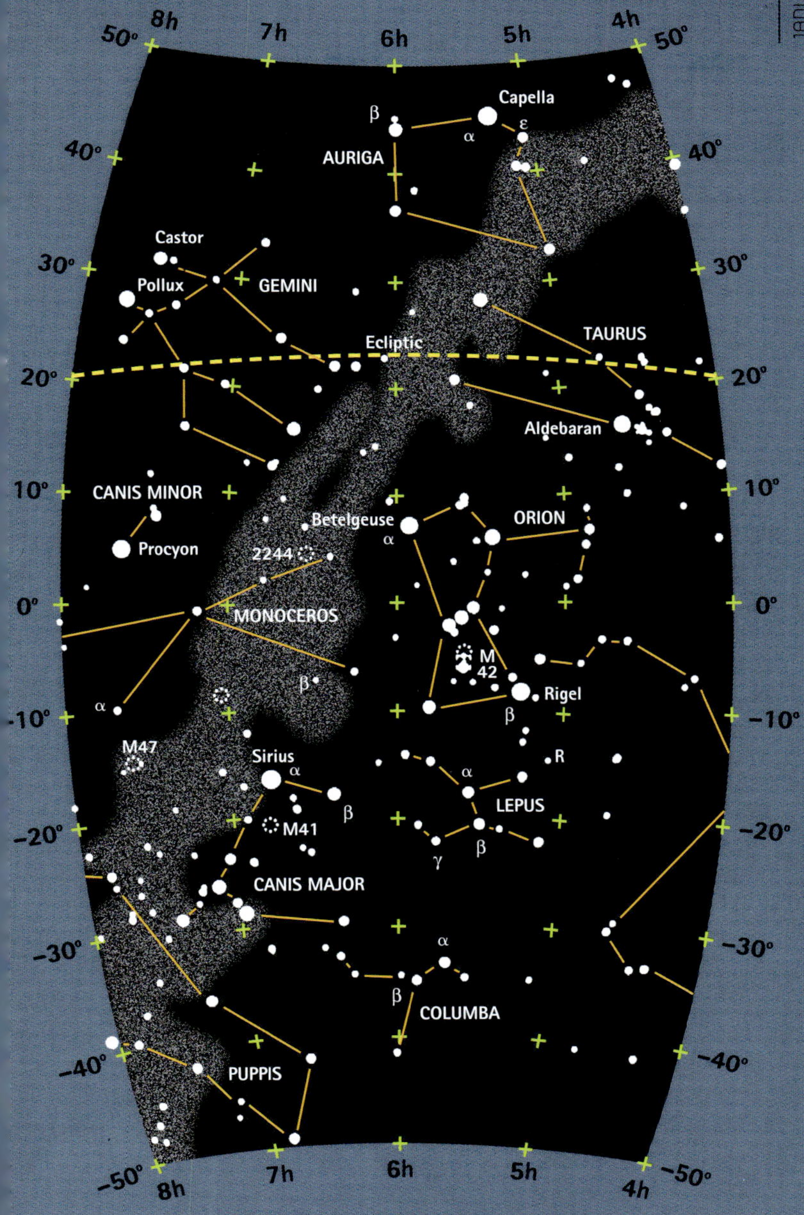
8h
7h
6h
5h
4h
50°
40°
30°
20°
10°
0°
-10°
-20°
-30°
-40°
-50°
Capella
AURIGA
Castor
Pollux
GEMINI
Ecliptic
TAURUS
Aldebaran
CANIS MINOR
Procyon
Betelgeuse
ORION
2244
MONOCEROS
M 42
Rigel
M47
Sirius
R
LEPUS
M41
CANIS MAJOR
COLUMBA
PUPPIS

COLUMBA, THE DOVE

One of the many bird constellations found in the far Southern Hemisphere, Columba represents the dove Noah released from the Ark to search for dry land. It is a small and relatively faint constellation, but is quite easy to spot because it appears in an otherwise featureless region of the sky.

A dark horse ▲ seemingly rears its head in one of Orion's gas clouds, creating the unmistakable Horsehead Nebula.

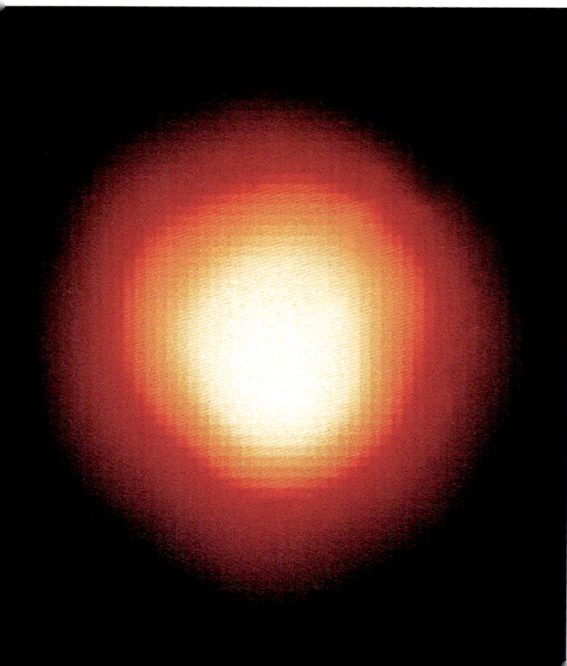

◄ The supergiant star Betelgeuse at Orion's right shoulder. The Hubble Space Telescope is so powerful that it has been able to show the star as a disc—a remarkable feat.

LEPUS, THE HARE

The Hare, which lies beneath the feet of Orion, is running away from the largest of his dogs, Canis Major. The star Gamma (γ) is a double, which can be seen through binoculars. One of the most interesting stars is R. It is a variable, which at its brightest can easily be seen through binoculars. It has a striking red color, and is often called the Crimson Star.

ORION

Straddling the celestial equator, this magnificent star group is this month's key constellation (see page 26).

The glorious Orion Nebula looks magnificent in telescopes. ▼

Orion

The Greeks named this magnificent constellation after a mighty hunter. He was the son of the sea god Poseidon, and boasted he could kill any creature on Earth. He strides across the heavens with a club raised in his right hand and with a shield in the other. A sword dangles from his belt. At his feet are his two dogs, Canis Major and Minor.

It is said that Orion fell in love with seven sisters, the Pleiades. And in the sky he still pursues them, for Orion follows the Pleiades star cluster as the heavens rotate. Orion is located in the opposite part of the sky to Scorpius, the Scorpion. The gods placed him there, it is said, to avoid confrontation, because it was the Scorpion that stung Orion to death.

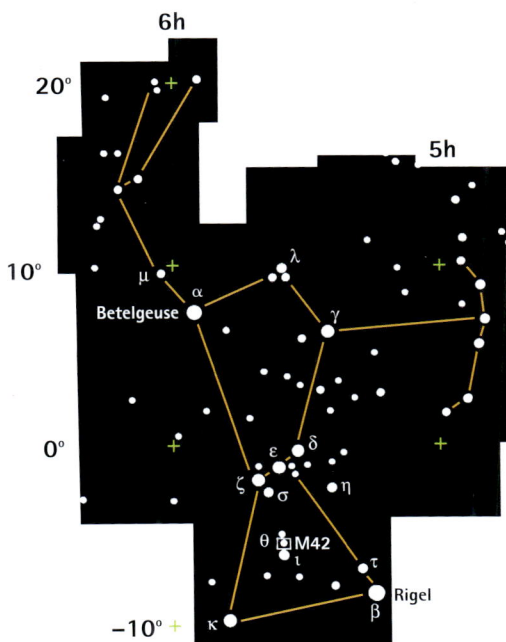

BIG AND BRIGHT

Orion is a familiar sight to stargazers of both hemispheres because it straddles the celestial equator. Its main outline is marked by seven bright stars. At the figure's left foot is Rigel, which is opposite Betelgeuse at Orion's right shoulder. The two make a contrasting pair: Rigel is brilliant white and slightly brighter than distinctly reddish Betelgeuse. Betelgeuse is a supergiant and one of the biggest stars we know, with an estimated diameter of at least 250 million miles (400 million km). If it were located where the Sun is in our solar system, it would reach out beyond Mars.

The other distinctive feature of the constellation figure is the group of three brilliant stars close together in the middle. They form Orion's belt. To their south is the glowing patch that forms part of Orion's sword. It is a vast region of glowing gas, known as the Great Nebula, the Orion Nebula, or M42. It is lit up by four stars in a multiple-star system called the Trapezium, embedded within it.

The Nebula is a hotbed of star formation, as are other nebulous regions in the constellations. One is located close to the most southerly star in Orion's belt (Zeta ζ). In this particular region a huge dark mass of gas is silhouetted against a bright background. It bears an uncanny resemblance to the head and neck of a horse, which is why it has been called the Horsehead Nebula.

NORTHERN HEMISPHERE

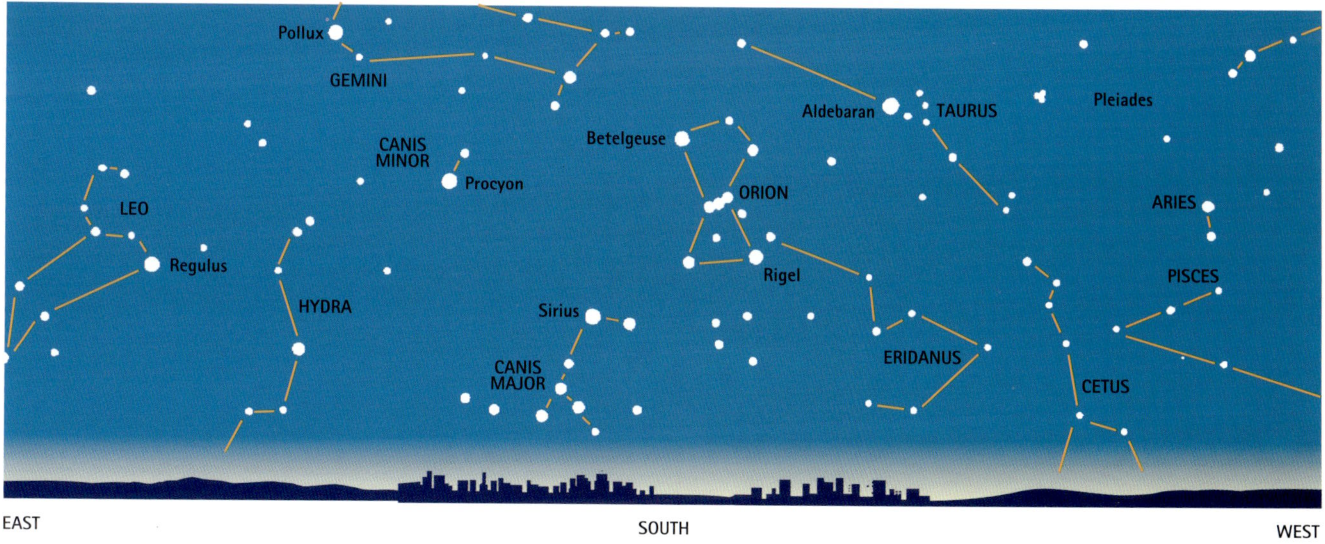

EAST

SOUTH

WEST

JANUARY SKIES—LOOKING SOUTH
Constellations visible in North America and Europe at about 11:00 P.M. on about January 7.

The unmistakable figure of Orion appears in mid-skies nearly due south, with orange Betelgeuse contrasting nicely with brilliant white Rigel. As ever, the mighty hunter serves as an incomparable signpost to a host of first-magnitude stars.

Southeast is brightest star in the sky Sirius, the Dog Star. Northwest is Aldebaran, the red eye of the charging bull, and beyond it, the magnificent Pleiades cluster. Northeast are the twins Castor and Pollux. Due east is Procyon.

NORTHERN HEMISPHERE

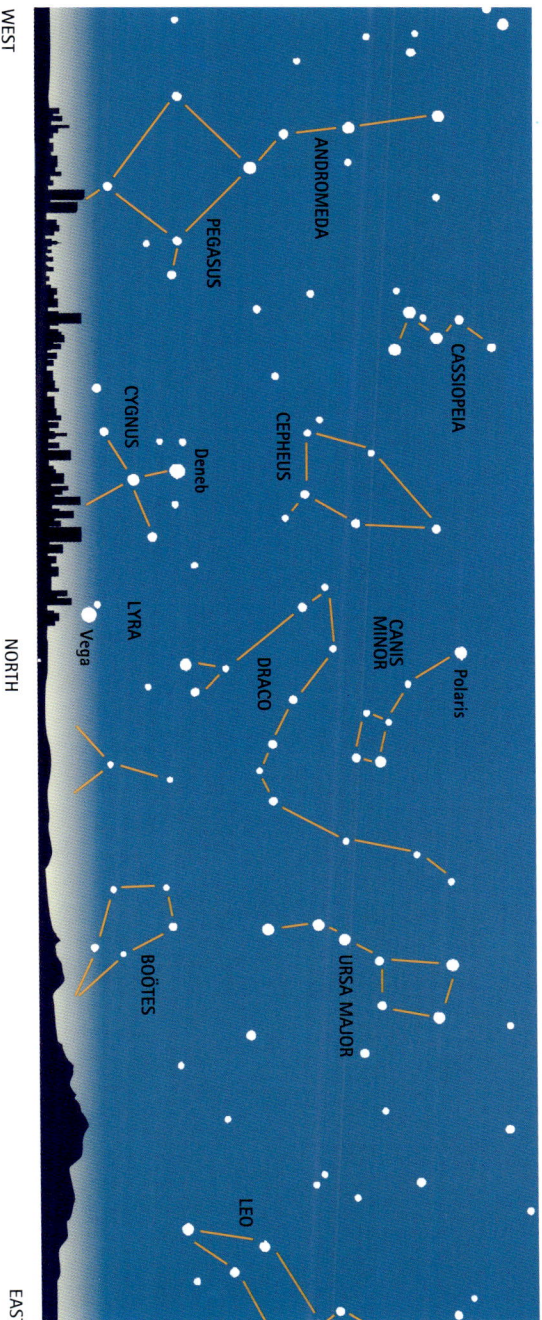

WEST

NORTH

EAST

JANUARY SKIES—LOOKING NORTH

Constellations visible in North America and Europe at about 11:00 P.M. on about January 7.

As in all the northern hemisphere views looking north, Polaris, the Pole or North Star, sits directly on the meridian, not noticeably deviating from its position during the year. Circling counterclockwise around it are the circumpolar constellations that stay visible all the time. In this January view from mid-latitudes, Cassiopeia, Cepheus, Draco, and Ursa Minor and Major (the Little and Big Dippers or Plow) are the circumpolar constellations. The others are higher up and overhead, out of the frame here.

SOUTHERN HEMISPHERE

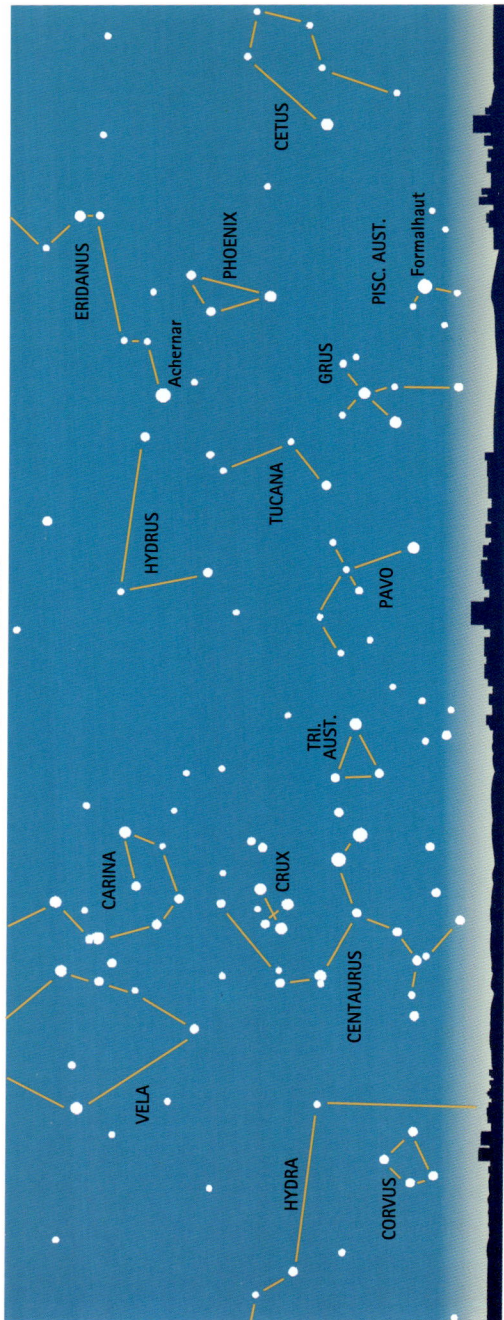

EAST

WEST

SOUTH

JANUARY SKIES—LOOKING SOUTH
Constellations visible in Australia and South Africa at about 11:00 P.M. on about January 7.

Unlike the northern hemisphere, the southern hemisphere has no convenient pole star. This is mildly disorientating and inconvenient for southern observers. But the long axis of Crux, the Southern Cross, acts as a passable pointer to the southern celestial pole. Around this pole, the southern stars rotate in a clockwise direction. Crux is one of the circumpolar constellations that is always above the horizon, along with its brilliant pointers, Alpha and Beta Centauri. So are Carina and Triangulum Australe.

SOUTHERN HEMISPHERE

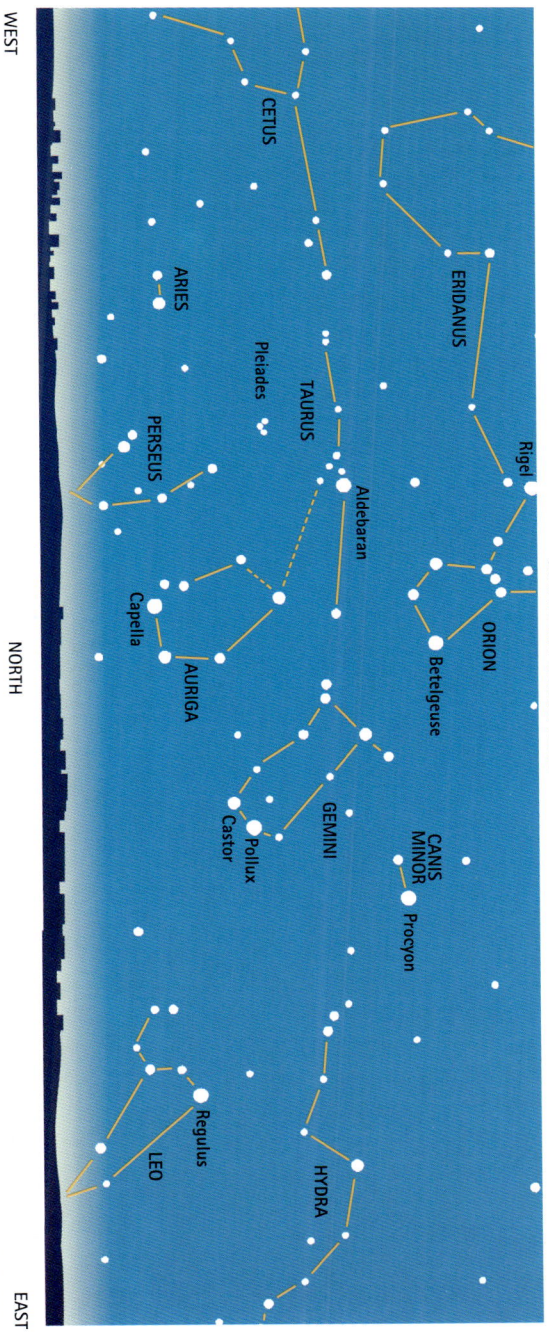

WEST   NORTH   EAST

JANUARY SKIES—LOOKING NORTH
Constellations visible in Australia and South Africa at about 11:00 P.M. on about January 7.

The unmistakable figure of Orion appears high in the sky nearly due north, with orange Betelgeuse contrasting nicely with brilliant white Rigel! As ever, the mighty hunter serves as an incomparable signpost to a host of first-magnitude stars. Due east of Betelgeuse is Procyon. Northeast are the twins Castor and Pollux. Low down near the northern horizon is brilliant Capella. Northwest is Aldebaran, the red eye of the charging bull, and beyond it, the magnificent Pleiades cluster.

February Stars

February skies are not as impressive as those of the previous month, since rather faint constellations are moving in from the east. Only in far southern skies do the Milky Way and bright constellations still dazzle.

CANCER, THE CRAB

This is one of the constellations of the zodiac, located between Gemini and Leo. The Sun passes through Cancer between July 20 and August 10. Its stars are relatively faint—in fact, it is the faintest constellation of the zodiac. Nevertheless, it has several interesting features.

The most impressive one is the Beehive, a fine open cluster visible to the naked eye but better seen with binoculars. It was given this name because ancient astronomers likened the clusters of stars to bees buzzing busily around their hive. Its proper astronomical name is Praesepe or M44.

Small telescopes will reveal that the star Iota (ι) is a double. The pair have lovely colors—gold and pale blue.

CANIS MINOR, THE LITTLE DOG

This tiny constellation represents the smaller of the two dogs that accompanied Orion when he went hunting. The constellation has only two bright stars, the brighter being named Procyon, which means something like "before the dog." This refers to the fact that it rises above the horizon before the Dog Star, Sirius. Procyon is the eighth-brightest star in the whole heavens, and one of the nearest, at a distance of about 11 light-years. Among bright stars, only Alpha Centauri and Sirius are closer. Procyon imitates Sirius in having a faint white dwarf companion.

GEMINI, THE TWINS

This fine constellation is found in the zodiac between Cancer and Taurus. The Sun passes through it between June 21 and July 20. Its pair of bright main stars, both of the first magnitude, are named Castor and Pollux. To the ancient Greeks, they represented the twin sons, hatched from an egg laid by Leda, Queen of Sparta, after she had been wooed by Zeus (see page 92). To the ancient Romans, they represented Romulus and Remus, the legendary founders of Rome, who were suckled and brought up by a wolf.

FEBRUARY SKIES ▶
Constellations visible near the meridian at about
11:00 P.M. during the first week in February.

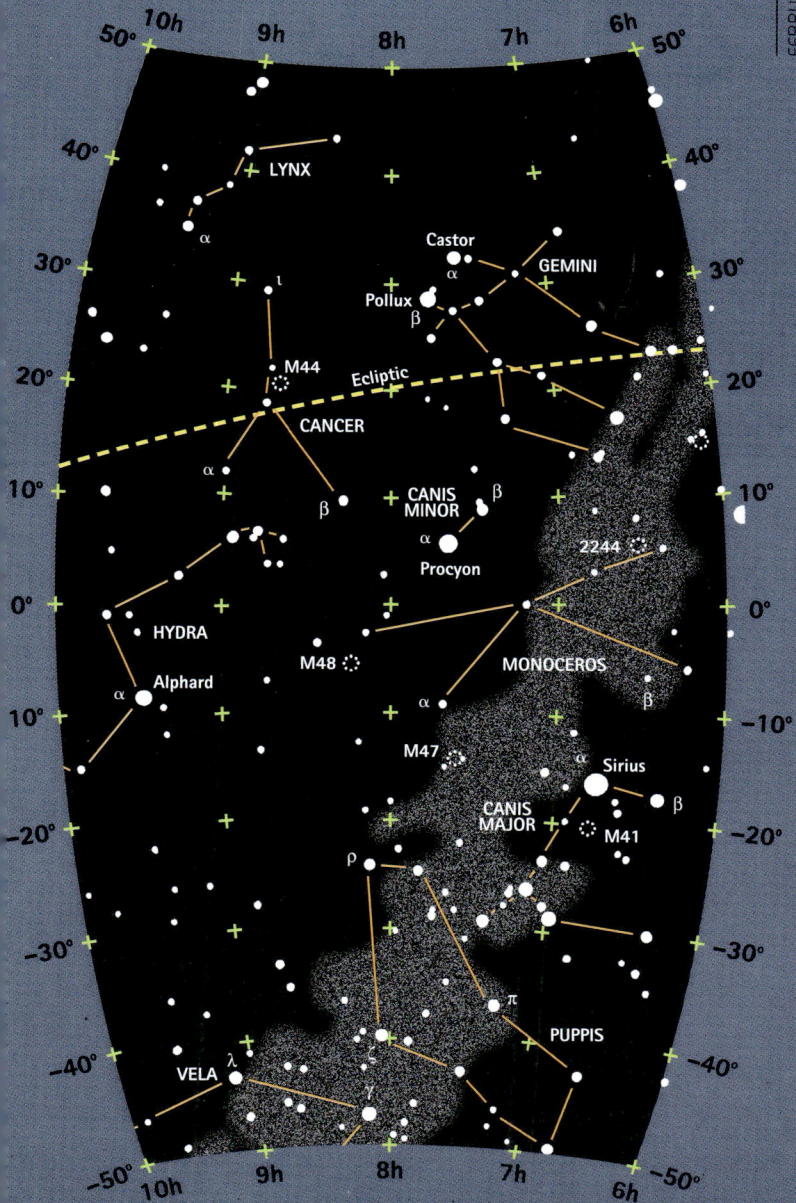

10h **9h** **8h** **7h** **6h**

50° 50°
40° 40°
30° 30°
20° 20°
10° 10°
0° 0°
–10° –10°
–20° –20°
–30° –30°
–40° –40°
–50° –50°

LYNX

α

ι

Castor
α
GEMINI
Pollux
β

M44
Ecliptic
CANCER

α

β

CANIS
MINOR
α
Procyon

β

2244

HYDRA

M48

MONOCEROS

β

α

Alphard

α

M47

α
Sirius

CANIS
MAJOR
M41

β

ρ

π

PUPPIS

VELA
λ
ι
γ

10h **9h** **8h** **7h** **6h**

Of the twins, Castor is slightly fainter, but is the more interesting. It is an impressive multiple-star system. Small telescopes will reveal that it is a binary system, consisting of two blue-white stars circling around each other. Examining their light through a spectroscope reveals that each one is also a binary, formed by a pair of close-orbiting stars. There is yet another spectroscopic binary in the system, making six stars in all. Astronomers find that multiple systems of stars like this are not at all unusual in the heavens.

MONOCEROS, THE UNICORN

This constellation has only relatively faint stars, but straddles a particularly rich region of the Milky Way. Its most interesting star is Beta (β). It is a multiple-star system, with three components, easily seen as a triangle in a small telescope. But the crowning glory of the constellation is a magnificent nebula aptly known as the Rosette. It is a circular cloud of red gas, with petal-like regions surrounding a cluster of brilliant stars (NGC2244).

PUPPIS, THE POOP OR THE STERN

Puppis and the neighboring constellations Vela and Carina were once included in a much larger constellation called Argo Navis. It represented the ship in which the Argonauts sailed to search for the Golden Fleece in a famous story of Greek mythology. Puppis represents the stern of the ship, Vela the sails, and Carina the keel.

Much of this brilliant southern constellation lies within the Milky Way, so it is rich in star clusters and nebulae. Brightest of the clusters is M47, easily visible to the naked eye in the Milky Way to the east of Sirius.

Clouds of glowing gas fill much of Monoceros, extending into ▼ neighboring Orion.

One of the gems in Monoceros, the ► lovely Rosette Nebula, with petal-like skeins of gas.

NORTHERN HEMISPHERE

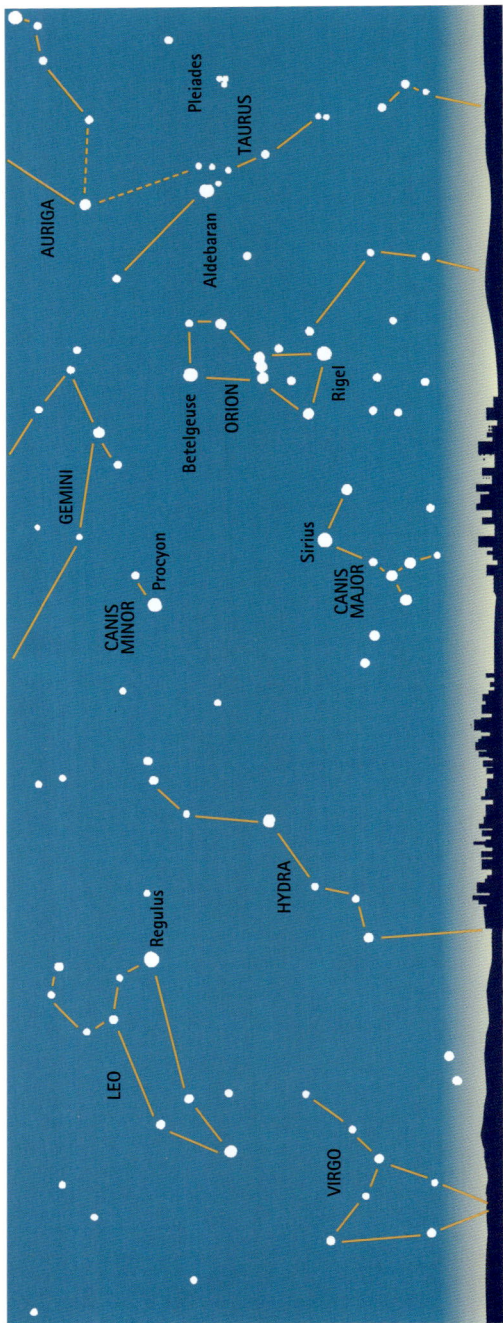

EAST

SOUTH

WEST

FEBRUARY SKIES—LOOKING SOUTH

Constellations visible in North America and Europe at about 11:00 P.M. on about February 7.

Mid-skies have lost some of their spectacle, with Orion and Taurus slipping away west, and with them the pearly white band of the Milky Way. But the two constellations are still well placed for observation. Canis Minor and its brightest star Procyon are now close to the meridian. Castor and Pollux are high above them, while Sirius shines brilliantly closer to the horizon. In lower northern latitudes, observers may be able to glimpse on the horizon the far southern constellations Puppis and Vela.

NORTHERN HEMISPHERE

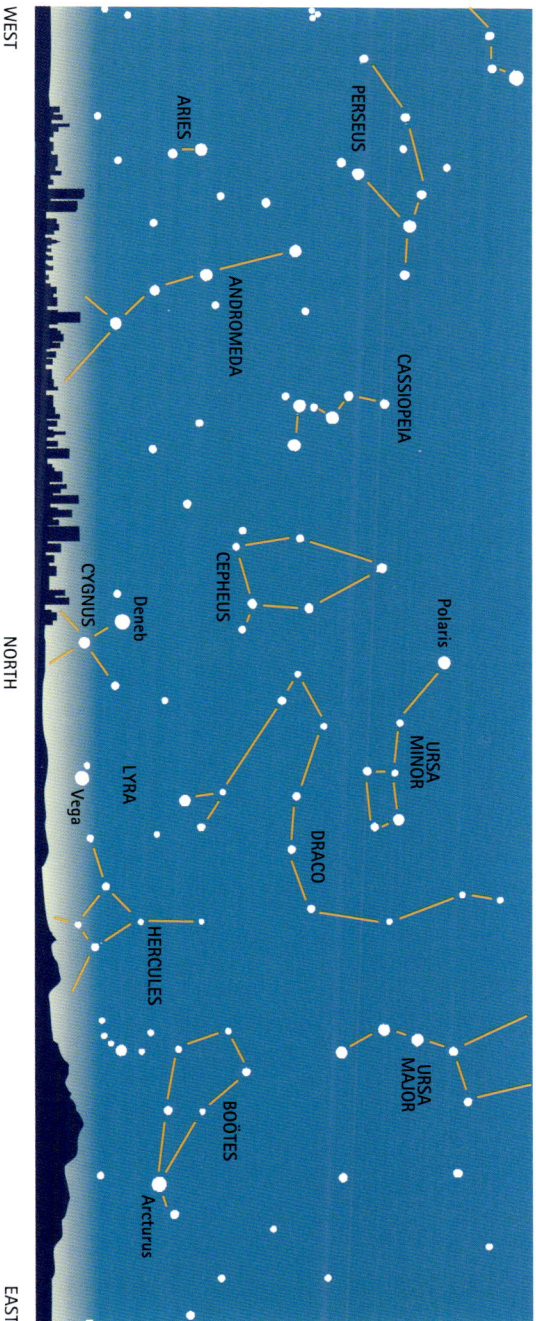

WEST

NORTH

EAST

FEBRUARY SKIES—LOOKING NORTH

Constellations visible in North America and Europe at about 11:00 P.M. on about February 7.

This month Cassiopeia and Cepheus are sinking lower, while the Big Dipper or Plow, on the opposite side of Polaris from Cassiopeia, is climbing higher. On the meridian, low down on the horizon, is bright Deneb, marking the tail of the swan (Cygnus). Just to the east is the somewhat brighter Vega. Arcturus becomes visible as Boötes climbs higher. West of the meridian, Andromeda is still visible below the unmistakable W shape of Cassiopeia. Keen eyes should be able to spot the misty patch of the Andromeda Galaxy.

SOUTHERN HEMISPHERE

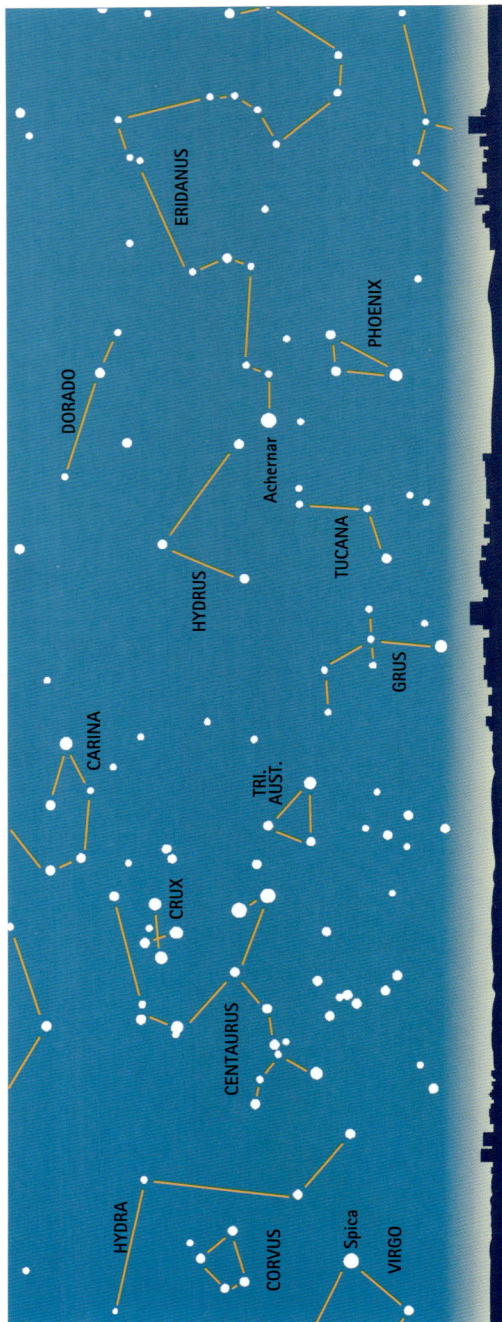

EAST

WEST

SOUTH

FEBRUARY SKIES—LOOKING SOUTH
Constellations visible in Australia and South Africa at about 11:00 P.M. on about February 7.

In February, the brightest part of the southern sky lies in the southeast, where the Southern Cross is located, its long axis nearly horizontal. As ever, it is encircled by the bright stars of Centaurus. In the far east, Spica, the lead star of Virgo, has risen. In the western half of the sky, only Achernar is conspicuous. It marks the mouth of the widely meandering and immensely long river Eridanus, whose mainly faint stars occupy much of the western sky.

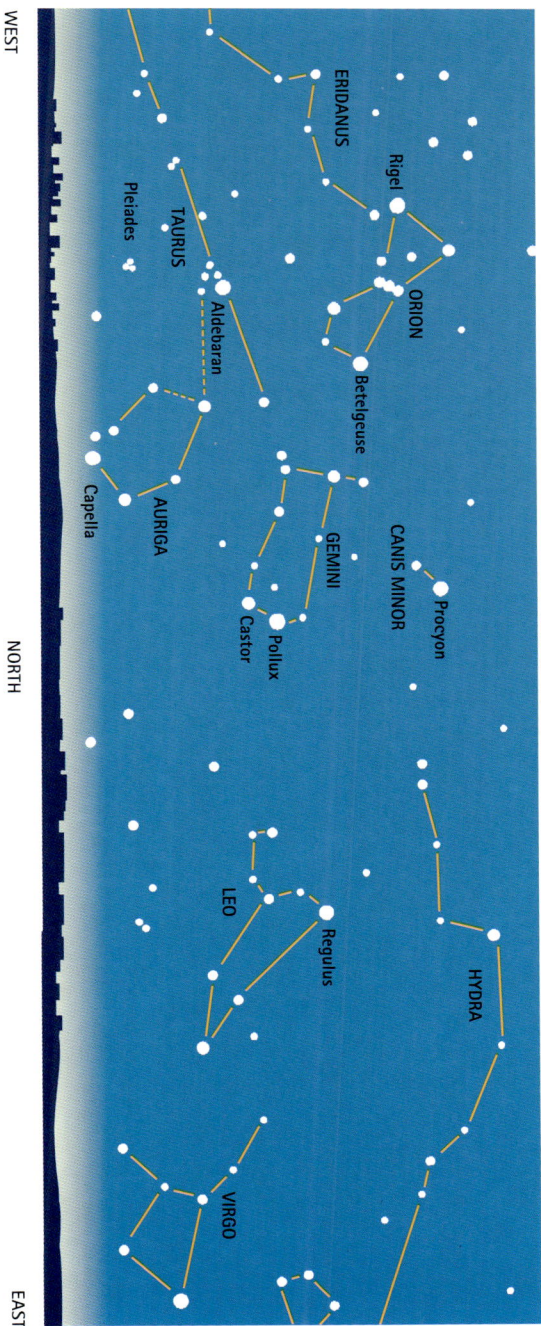

SOUTHERN HEMISPHERE

WEST

NORTH

EAST

FEBRUARY SKIES—LOOKING NORTH
Constellations visible in Australia and South Africa at about 11:00 P.M. on about February 7.

Mid-skies have lost some of their spectacle, with Orion and Taurus slipping away west, and with them the pearly white band of the Milky Way. But the two constellations are still well placed for observation. Canis Minor and its brightest star

Procyon are now close to the meridian. Castor and Pollux are in mid-skies below them. Northwest near the horizon, Capella is still just visible, as are the Pleiades farther west still.

March Stars

March skies are relatively bare and lack the starry arch of the Milky Way. But they are saved from complete obscurity by the zodiac constellation Leo, with its easy-to-recognize Sickle. When Leo appears in the sky, the days are lengthening and spring is coming to the Northern Hemisphere. In the Southern, it is the nights that are lengthening and the fall that is approaching.

HYDRA, THE WATER SNAKE (HEAD)

Hydra stretches a quarter of the way around the celestial sphere and is the largest of the 88 constellations. Its undulating, serpentlike line of stars runs roughly parallel with, and to the south of, the ecliptic and the zodiac constellations of Libra, Virgo, Leo, and Cancer. In Greek mythology Hydra was the multiheaded sea snake that Hercules wrestled with and finally killed as one of his 12 labors. The ancient Egyptians, however, likened the constellation to their long and life-giving river, the Nile.

Hydra may be long, but its stars are far from spectacular. It has only one star that can be considered bright, the second-magnitude Alphard. This is a noticeably orange-colored giant. The name Alphard means the "solitary one," an apt name, for it appears in a relatively bare region of the heavens. It is easily found by dropping south from the bright Regulus in the unmistakable Leo.

Among the few other features of interest in the constellation, the bunch of stars that form the snake's head look lovely in binoculars. So does the star cluster M48, located south of these head stars.

See page 51 for Hydra (tail).

LEO, THE LION

This constellation of the zodiac is formed by reasonably bright stars. The Sun passes through the constellation between August 10 and September 16 every year. Leo is quite a good likeness to the figure of a crouching lion, poised and ready to pounce on its unsuspecting prey. The curved line of stars that mark the lion's head and front is known as the Sickle since it describes the distinctive shape of a garden sickle. In mythology, the lion was another victim of Hercules.

The lead star of the constellation is Regulus. It is also known as Cor Leonis, which means the Lion's Heart. It is a double star, as is another

MARCH SKIES ▶
Constellations visible near the meridian at about
11:00 P.M. during the first week in March.

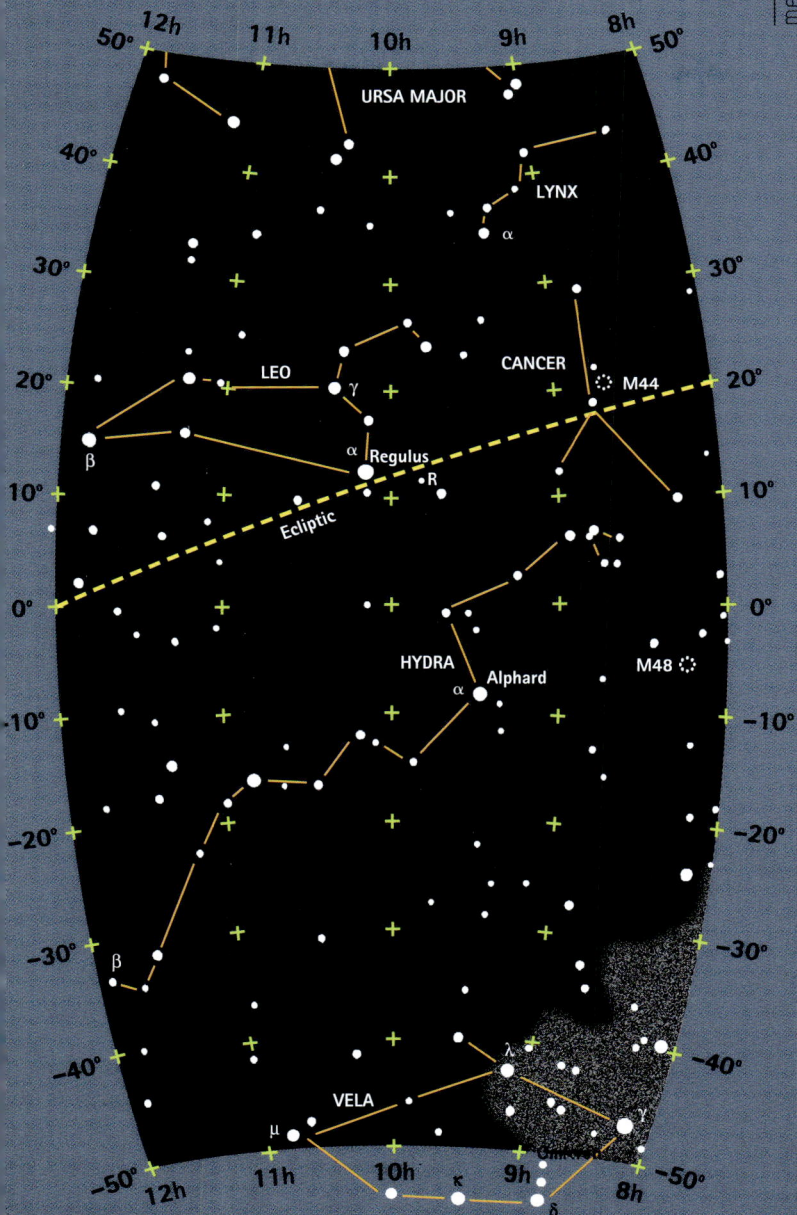

12h **11h** **10h** **9h** **8h**
50° 50°

URSA MAJOR

40° 40°

LYNX

α

30° 30°

LEO CANCER

20° γ ⊙ M44 20°

α Regulus
β • R

10° 10°
Ecliptic

HYDRA

0° 0°

Alphard
α M48 ⊙

-10° -10°

-20° -20°

β

-30° -30°

λ

VELA γ

-40° -40°

μ
κ
-50° δ -50°
12h **11h** **10h** **9h** **8h**

Sickle star, Gamma (γ), which is also called Algeiba, or the Lion's Mane. Still at the front end, there is the reddish Mira variable star R, which brightens to the fourth magnitude and fades to the eleventh magnitude over a period of about 10 months.

The tail end of the lion is marked by a triangle of bright stars. Beta (β), or Denebola, is a fine binary, with golden yellow components. It looks lovely in a small telescope. The tail end of the lion merges into Virgo and into a region that is rich in galaxies. Two pairs of spirals can be spotted in small telescopes south of the Denebola–Regulus line of stars.

On about November 17 or 18 each year Leo plays host to a meteor shower, which can reach storm proportions in favorable years. The radiant—the point from which they appear to come—lies in the Sickle. The source of the shower is Comet Temple-Tuttle, which returns to the vicinity of the Sun every 33 years. As a result, showers peak over the same 33-year period. In 1966, Leonid meteors came thick and fast, peaking at a rate of 1,000 a minute; 1999 was also a good year.

◀ A fine ancient print of Leo, and its sidekick Leo Minor, the Little Lion. We can see that Regulus marks the big lion's heart.

A stunning vista in Vela, where the Milky Way is dense. ▲

LYNX, THE LYNX

This is a relatively recent (1600s) constellation, first described by the Polish astronomer Hevelius. He chose that name as he felt that only the lynx-eyed (with sharp vision) would be able to spot it. And it is faint, having only one star, Alpha (α), reasonably bright, at the third magnitude. The most interesting object in the constellation is way beyond the power of amateur telescopes. It is the globular cluster NGC2419. It lies at a distance of more than 200,000 light-years, making it further away than the nearest galaxy, the Large Magellanic Cloud! It has earned its name of the Intergalactic Wanderer.

VELA, THE SAILS

Once part of the large ancient Greek constellation Argo Navis (see page 34), Vela is one of the splendid far southern constellations that northern astronomers can never see. It is embedded in the Milky Way and is rich in clusters and nebulae. Gamma (γ) is a multiple star that reveals up to four components in a small telescope. IC2391, around Omicron, north of Delta (δ), is an open cluster that can be seen with the naked eye, but looks much better in binoculars. Delta and Kappa (κ) are two of the stars that make up the False Cross, along with two in Carina (page 18).

NORTHERN HEMISPHERE

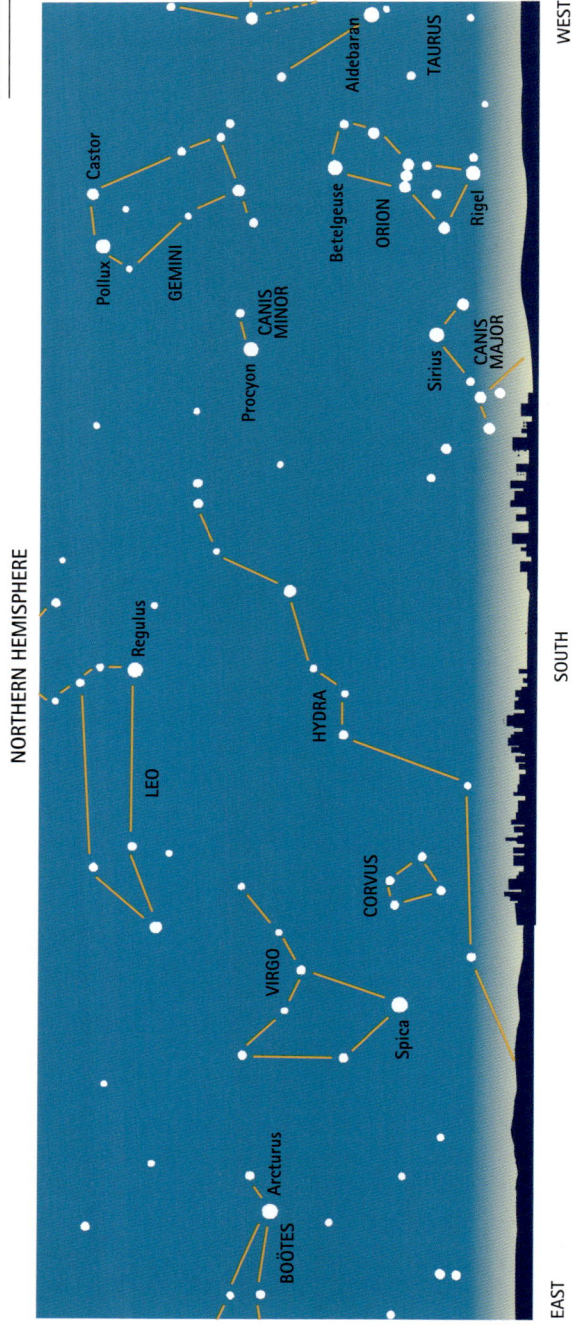

WEST

SOUTH

EAST

MARCH SKIES—LOOKING SOUTH

Constellations visible in North America and Europe at about 11:00 P.M. on about March 7.

Orion and Canis Major are sinking low on the western horizon, and Rigel and Sirius will soon be setting. Mid-skies are occupied by the serpentlike line of faint stars of Hydra, but higher up Leo crouches ready to pounce. Its brightest star

Regulus is located due south, with the sicklelike curve of stars above it. Virgo's lead star Spica shines brightly in the southeast, making with nearby Arcturus a prominent pair.

NORTHERN HEMISPHERE

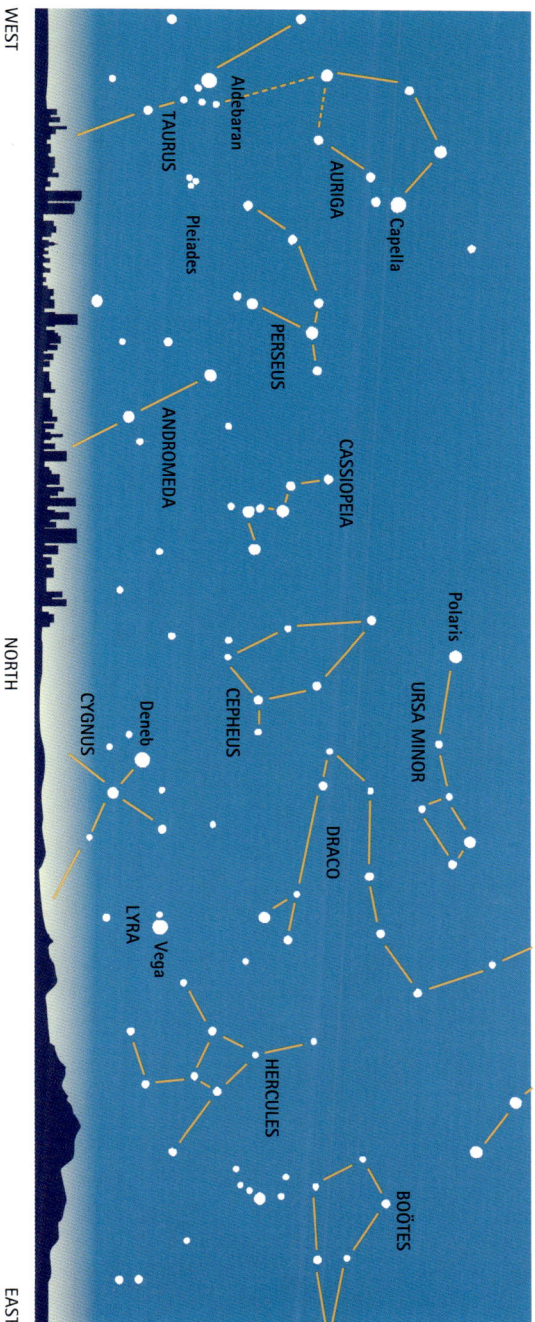

WEST

NORTH

EAST

MARCH SKIES—LOOKING NORTH

Constellations visible in North America and Europe at about 11:00 P.M. on about March 7.

Cepheus is on the meridian, directly beneath Polaris. To the east and lower down are the bright pair Deneb and Vega. They are two of the three stars (the other is Altair) that will in a few months time form the celebrated Summer Triangle. Continuing east, Hercules and Boötes are climbing,

and sandwiched between them is the arc of stars that form the Northern Crown, Corona Borealis. On the other side of the sky, Taurus and Auriga are descending, their brilliant stars Aldebaran and Capella shining like beacons.

SOUTHERN HEMISPHERE

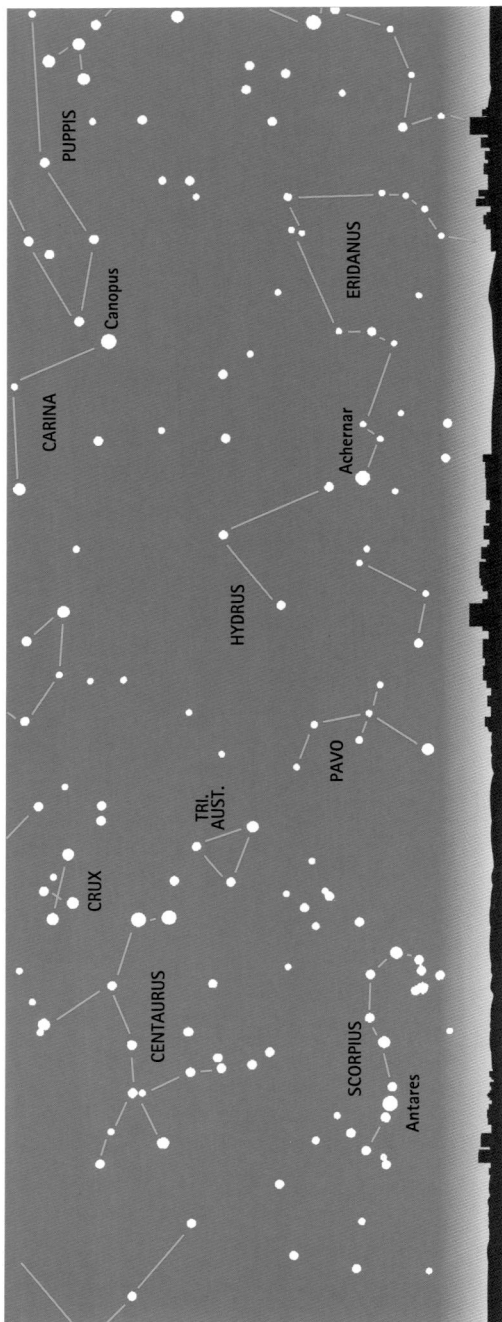

EAST

SOUTH

WEST

PUPPIS

CARINA

Canopus

HYDRUS

ERIDANUS

Achernar

CRUX

TRI. AUST.

PAVO

CENTAURUS

SCORPIUS

Antares

MARCH SKIES—LOOKING SOUTH

Constellations visible in Australia and South Africa at about 11:00 P.M. on about March 7.

As ever, center skies are empty of really bright stars around where the celestial south pole is located. But there are plenty in the southeast again. Crux and Centaurus dazzle, with Alpha and Beta Centauri now one above the other. Scorpius has risen above the eastern horizon, with Antares, marking the scorpion's heart, a noticeable orange color. The Milky Way here is sumptuously rich. Across the sky in the southwest, Canopus is unmistakable: not only is it the second brightest of all stars, but it appears in an otherwise bland region of the sky away from the Milky Way.

SOUTHERN HEMISPHERE

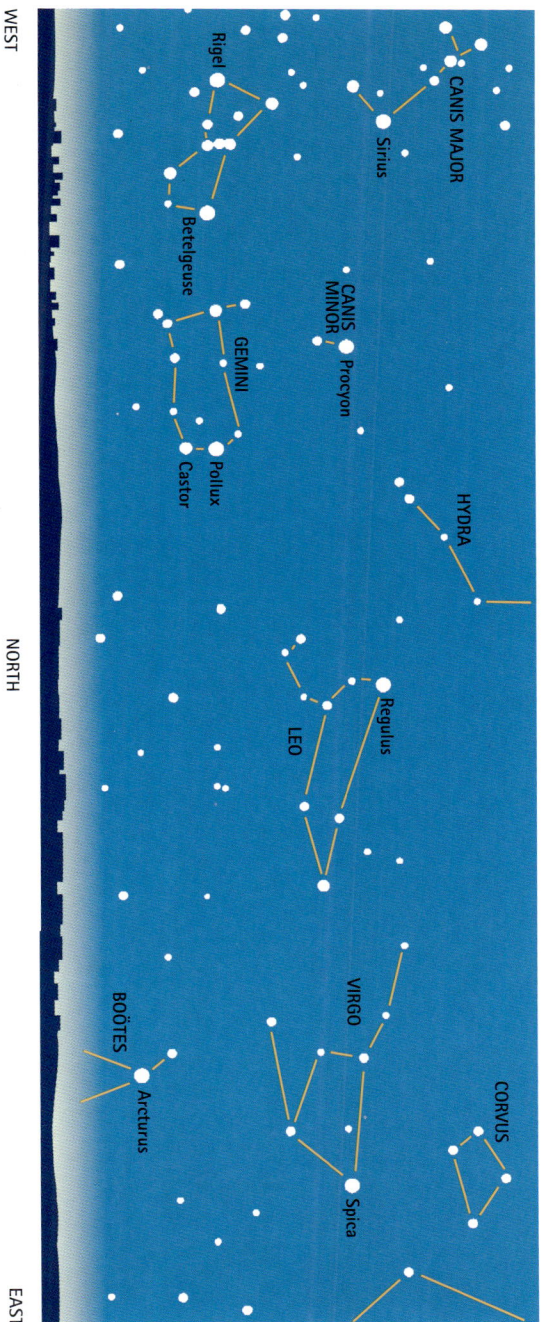

WEST

NORTH

EAST

MARCH SKIES—LOOKING NORTH

Constellations visible in Australia and South Africa at about 11:00 P.M. on about March 7.

Orion is sinking low on the western horizon, and the brilliant Rigel and Betelgeuse will soon be setting. They form a circle of bright stars with Castor and Pollux in nearby Gemini and Procyon higher up. Leo is prominent in mid-skies, crouching ready

to pounce. Its brightest star Regulus is located due north, with the sicklelike curve of stars below it. Virgo's lead star Spica shines brightly at about the same height in the east, making with Arcturus lower down a prominent pair.

April Stars

Leo lends some interest to the night sky, but to the south the skies are still relatively bare. This is because they are occupied by the two largest constellations—Virgo and Hydra—which possess few bright stars. Virgo at least has first-magnitude Spica, which forms a noticeable spring/fall triangle with two other equally bright stars—Regulus in Leo and Arcturus in Boötes.

CANES VENATICI, THE HUNTING DOGS

This constellation of mainly faint stars is disappointing to the naked eye, but is far more interesting in the telescope. It is a relatively modern constellation (1600s), whose subjects are supposed to be the dogs the herdsman Boötes used for hunting and protecting his herd, particularly from the attentions of the two bears (Ursa Major and Minor).

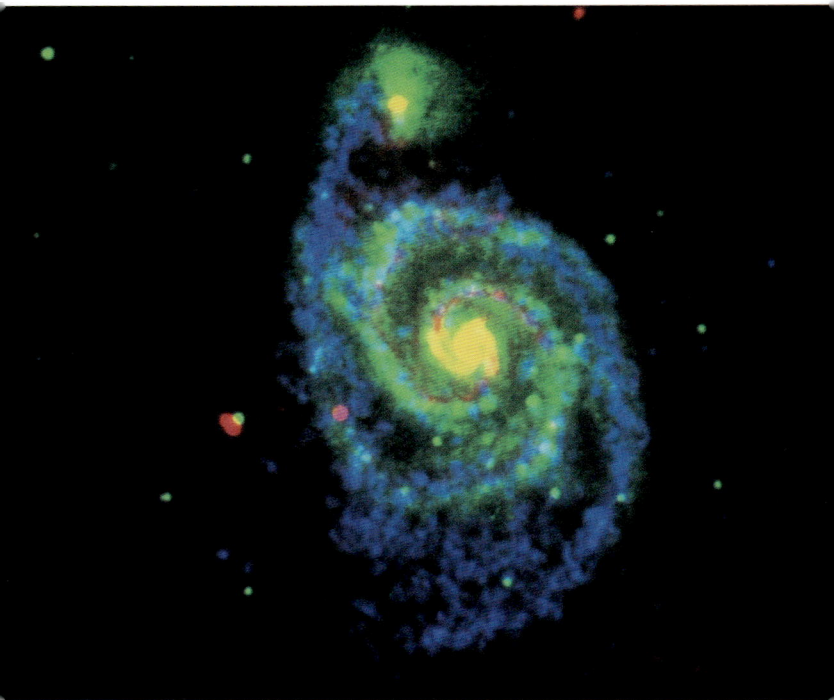

▲ The incomparable Whirlpool Galaxy, M51, a head-on spiral.

APRIL SKIES ▶
Constellations visible near the meridian at about 11:00 P.M. during the first week in April.

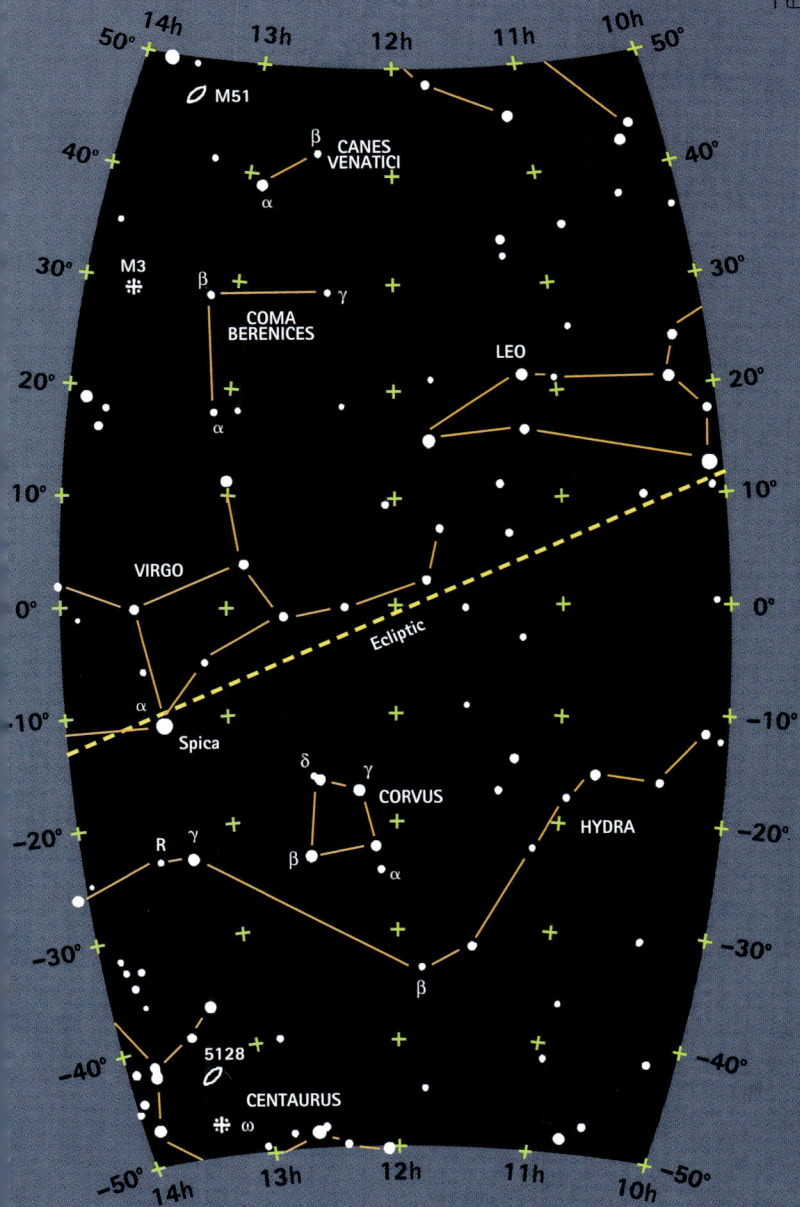

14h 13h 12h 11h 10h

50° 40° 50°

M51

β CANES
VENATICI
α

40° 40°

30° M3 β γ 30°
✳ COMA
BERENICES

LEO

20° 20°
α

α

10° 10°

VIRGO Ecliptic

0° 0°

α

−10° Spica −10°

δ γ
CORVUS

R γ HYDRA
−20° −20°
β α

β

−30° −30°

β

5128
−40° −40°
CENTAURUS
✳ ω

−50° −50°
14h 13h 12h 11h 10h

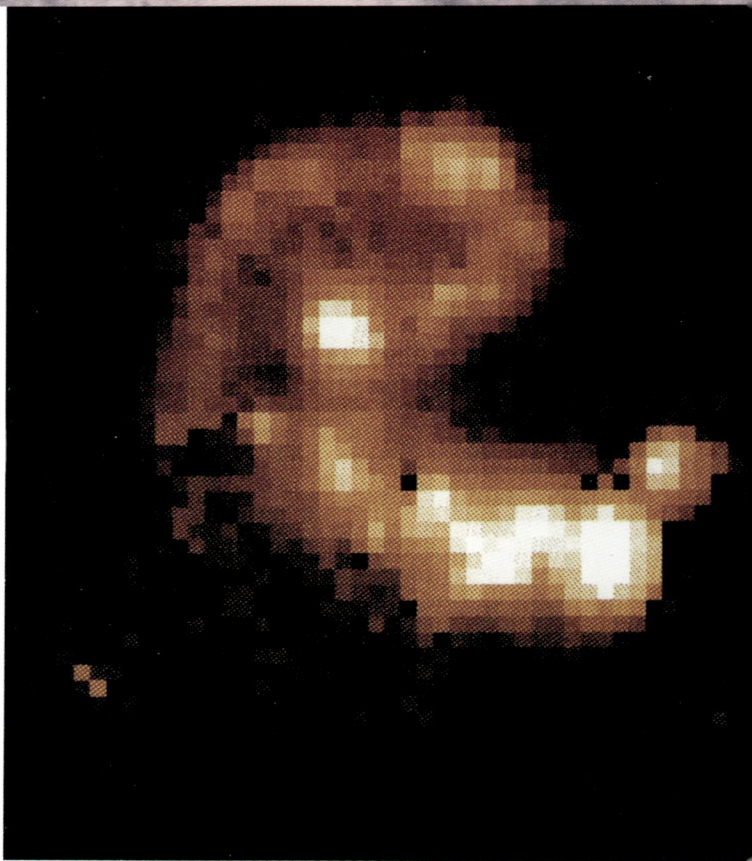

▲ An infrared image of the Antennae galaxies in Corvus.

Alpha (α) is a third-magnitude star, which in a small telescope proves to be a double. This star was given the name Cor Caroli, meaning Charles's Heart, by the English Astronomer Royal Edmond Halley (of the comet fame). It refers to the royal heart of the executed English king, Charles I.

There are two particularly interesting objects for telescopic observers. One is the Whirlpool Galaxy, M51. This is located on a line between Alpha and the first star in the handle of the Big Dipper (Plow). The Whirlpool is a particularly beautiful spiral galaxy that we see head on, with well-defined open spiral arms. Moreover, it is joined to, and interacts with, another smaller galaxy. Large telescopes spot many other galaxies in this region, which are on the edge of the vast Coma/Virgo cluster of galaxies.

The second highlight of Canes Venatici is M3, a fine globular cluster. This is found roughly halfway between Alpha and the bright Arcturus in the neighboring constellation Boötes. It can easily be spotted with binoculars, and telescopes will show it as an enormous globe of stars— hundreds of thousands of them.

CORVUS, THE CROW

This small constellation is linked in mythology with Hydra (the Water Snake), and Crater (the Cup). Apollo sent the crow to bring water to him in a cup. The crow stopped to eat figs, but blamed Hydra for being late. Apollo was not fooled, and sent the crow to the heavens, where, just out of reach of the cup, it is eternally thirsty.

Corvus is not a particularly interesting constellation, but is pleasing enough when viewed in binoculars. It contains several double stars, including Delta (δ). Large telescopes reveal near Gamma (γ) a pair of interacting galaxies, linked by curving skeins of glowing gas. They are collectively called the Antennae, because they look like the coiled antenna of a butterfly.

HYDRA, THE WATER SERPENT (TAIL)

Near Gamma (γ) is the reddish star R, which is a Mira-type variable. This means that it is a red giant that is easily visible with the naked eye at its brightest (magnitude 3), but fades to a difficult binocular object at its faintest (magnitude 11). It goes through this bright and faint cycle about every 13 months.

See page 40 for Hydra (head).

VIRGO, THE VIRGIN

Virgo is one of the constellations of the zodiac, sandwiched between Leo and Libra. The Sun passes through Virgo between September 16 and October 31 every year. Among the constellations, Virgo covers the second-largest area of the heavens, after Hydra. It is a sprawling constellation, not easy to make out in the sky.

Virgo has roots going back to Babylonian times, usually in the role of a mother goddess. In Greek mythology, she was goddess of justice and sometime corn goddess, and is usually depicted carrying a sheaf of wheat. This is reflected in the name of its one outstanding star, the first-magnitude Spica, which means "ear of wheat."

Only in large telescopes is Virgo really impressive. This is because it plays host to a dazzling array of distant galaxies. They occupy a broad band that runs through the whole western side of the constellation. These galaxies form part of one of the largest groupings of galaxies we know, called the Coma/Virgo cluster, which altogether contains at least 3,000 galaxies. It is one of the nearest clusters of galaxies, with its center only about 40 million light-years away.

NORTHERN HEMISPHERE

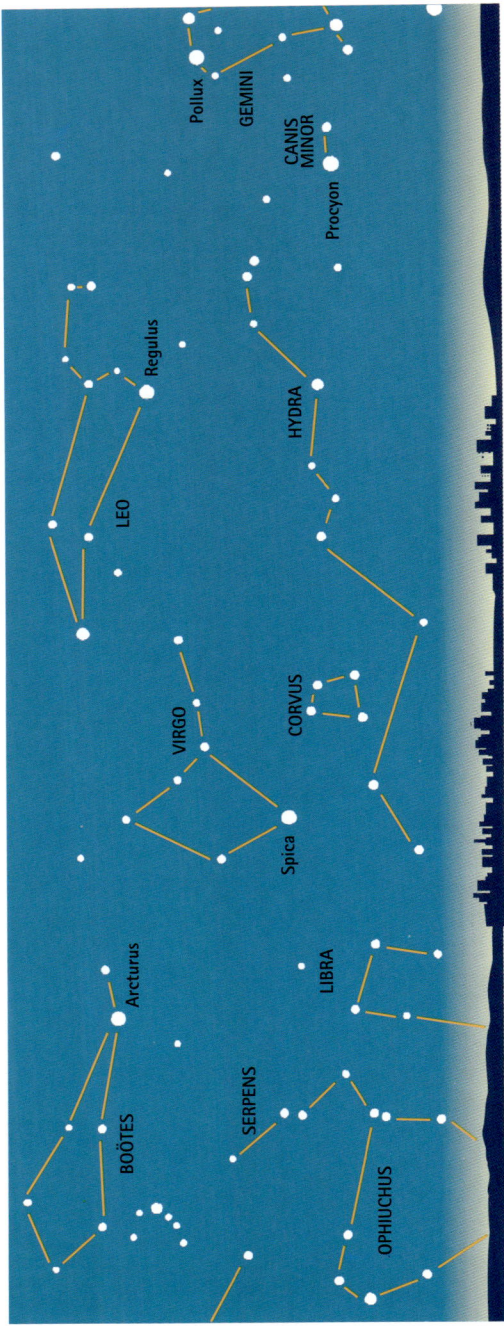

EAST

SOUTH

WEST

APRIL SKIES—LOOKING SOUTH

Constellations visible in North America and Europe at about 11:00 P.M. on about April 7.

Orion has all but disappeared below the horizon in the west, where Procyon, Castor and Pollux still shine. Leo is still prominent, just west of the meridian. Virgo now occupies center stage, but rather faintly. Only Spica stands out. High in the east Boötes is advancing, easily recognized by its distinctive kite shape and the brilliant Arcturus at its tail end. The rest of the sky, however, is disappointingly bland.

NORTHERN HEMISPHERE

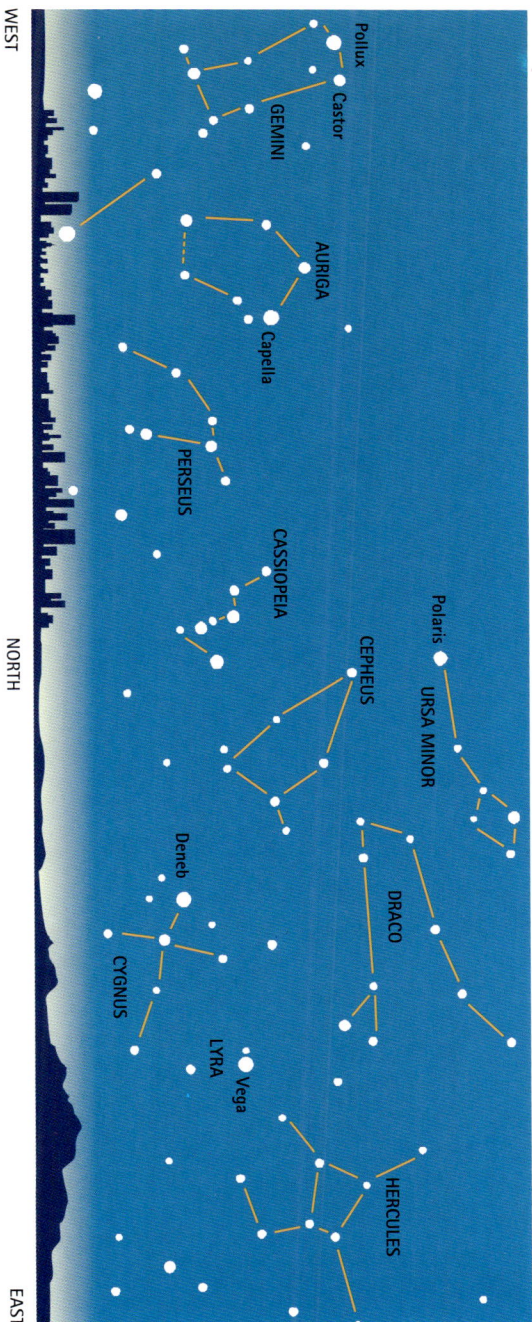

WEST

NORTH

EAST

APRIL SKIES—LOOKING NORTH
Constellations visible in North America and Europe at about 11:00 P.M. on about April 7.

Cassiopeia is reaching its low point in the northern heavens, almost on the meridian. On the opposite side of Polaris (and out of the frame here), the Big Dipper or Plow has climbed to its highest point, with its handle roughly parallel with the horizon.

In the east, Cygnus is still climbing and is now completely above the horizon. And we can fully appreciate the swanlike shape of its bright stars. The Milky Way is now almost parallel with the horizon, running from the feet of Gemini, through Cygnus.

SOUTHERN HEMISPHERE

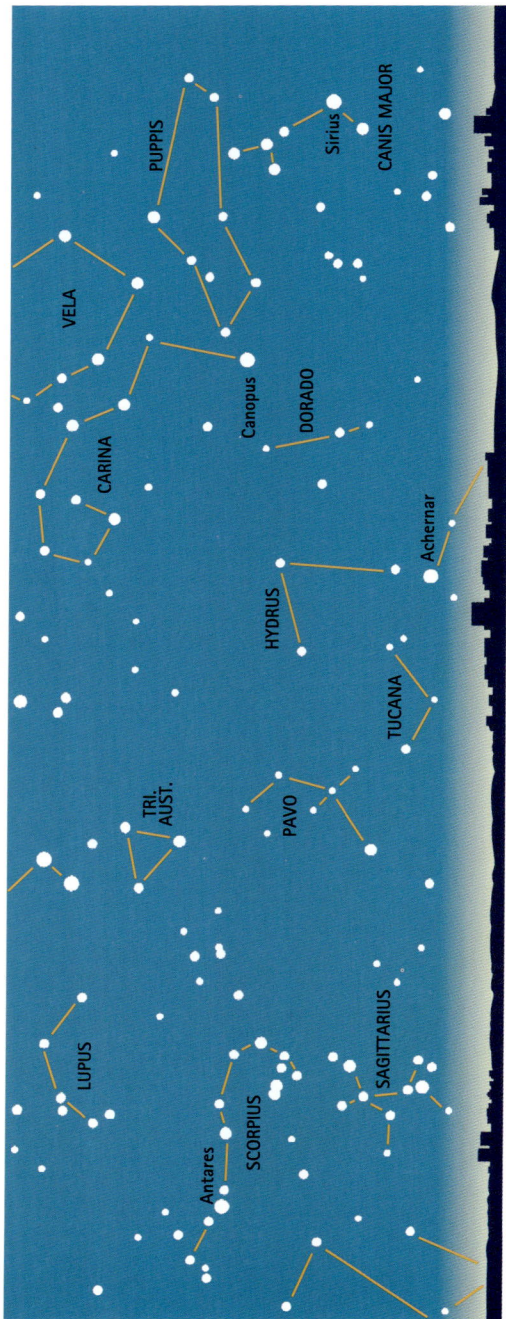

EAST

SOUTH

WEST

PUPPIS

CANIS MAJOR

Sirius

VELA

CARINA

Canopus

DORADO

HYDRUS

Achernar

TUCANA

PAVO

TRI. AUST.

LUPUS

SCORPIUS

Antares

SAGITTARIUS

APRIL SKIES—LOOKING SOUTH

Constellations visible in Australia and South Africa at about 11:00 P.M. on about April 7.

Crux has risen out of the frame here, but we know where it is by reference to the prominent pointers Alpha and Beta Centauri. It sits nearly on the meridian. Low down, the toucan (Tucana) has reached its lowest position in the sky, while its neighbor the peacock (Pavo) is ascending.

Achernar, too, is only just above the southern horizon. Scorpius is still brightening the skies in the east, while Sirius has joined Canopus in the west. It seems strange that the two brightest stars in the heavens should be so close together.

SOUTHERN HEMISPHERE

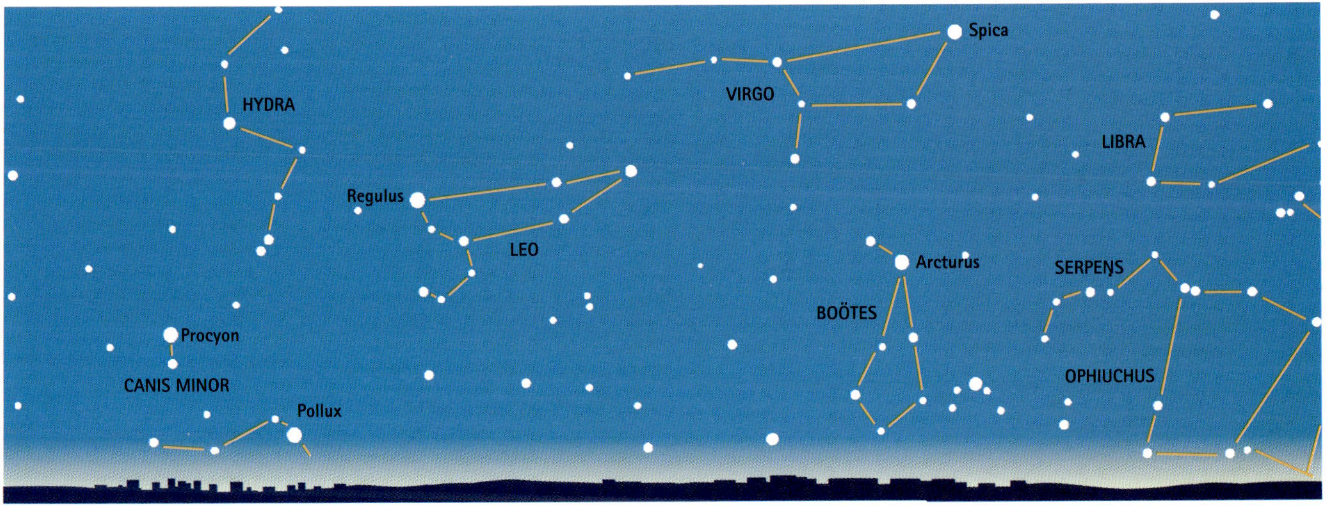
HYDRA
Regulus
LEO
VIRGO
Spica
LIBRA
Arcturus
BOÖTES
SERPENS
OPHIUCHUS
Procyon
CANIS MINOR
Pollux

WEST NORTH EAST

APRIL SKIES—LOOKING NORTH

Constellations visible in Australia and South Africa at about 11:00 P.M. on about April 7.

Orion has disappeared below the horizon in the west, and Pollux is the only Gemini twin still visible. Procyon, too, is close to setting. Leo is still prominent, as it drifts west. Virgo is close to the meridian, but not at all easy to make out. Only Spica stands out, almost directly above Arcturus at the tail end of the easy-to-recognize kite shape of Boötes. However, the rest of the sky is disappointingly bland, with Libra, Serpens and Ophiuchus risen in the east.

May Stars

The two bright stars that occupy the mid-skies make an interesting contrast. The northern one is Arcturus, a noticeably orange-red giant star. The slightly dimmer southern one is Spica, also a giant but pure white. Apart from these two stars, however, the mid-sky region remains bland, occupied by the faint stars of Serpens (head), Libra, Virgo, and the tail of Hydra.

BOÖTES, THE HERDSMAN

This prominent northern constellation is easy to recognize by its kite shape, with bright Arcturus marking the tail end of the kite. It is located to the south of Ursa Major, the Great Bear, and is easily located by following the curve of the stars in the handle of the Big Dipper (Plow).

In Greek mythology, Boötes was Arcas, son of Callisto, who was changed into a bear (Ursa Major). Out hunting one day, Boötes didn't recognize her and was about to kill her, but Zeus (Callisto's former lover) placed them both in the heavens to prevent this tragedy from occurring. The Greek poet Homer saw things differently, considering Boötes the Bear Driver, chasing the Great and Little Bears across the sky.

The bear theme is continued in the name of the constellation's lead star Arcturus. The name means "bear guard." Arcturus is a red giant star some 20 million miles (30 million km) across, about 20 times bigger than the Sun. Highly luminous and relatively close (25 light-years), it is the fourth-brightest star in the whole heavens.

Three of Boötes' bright stars are doubles, visible in a small telescope— Epsilon (ϵ), Xi (ξ), and Mu (μ), which binoculars will also separate. Epsilon is particularly lovely, with red and blue-green components.

▲ Most of the constellation of Centaurus, which wraps itself around the unmistakable Crux, the Southern Cross.

MAY SKIES ▶
Constellations visible near the meridian at about 11:00 P.M. during the first week in May.

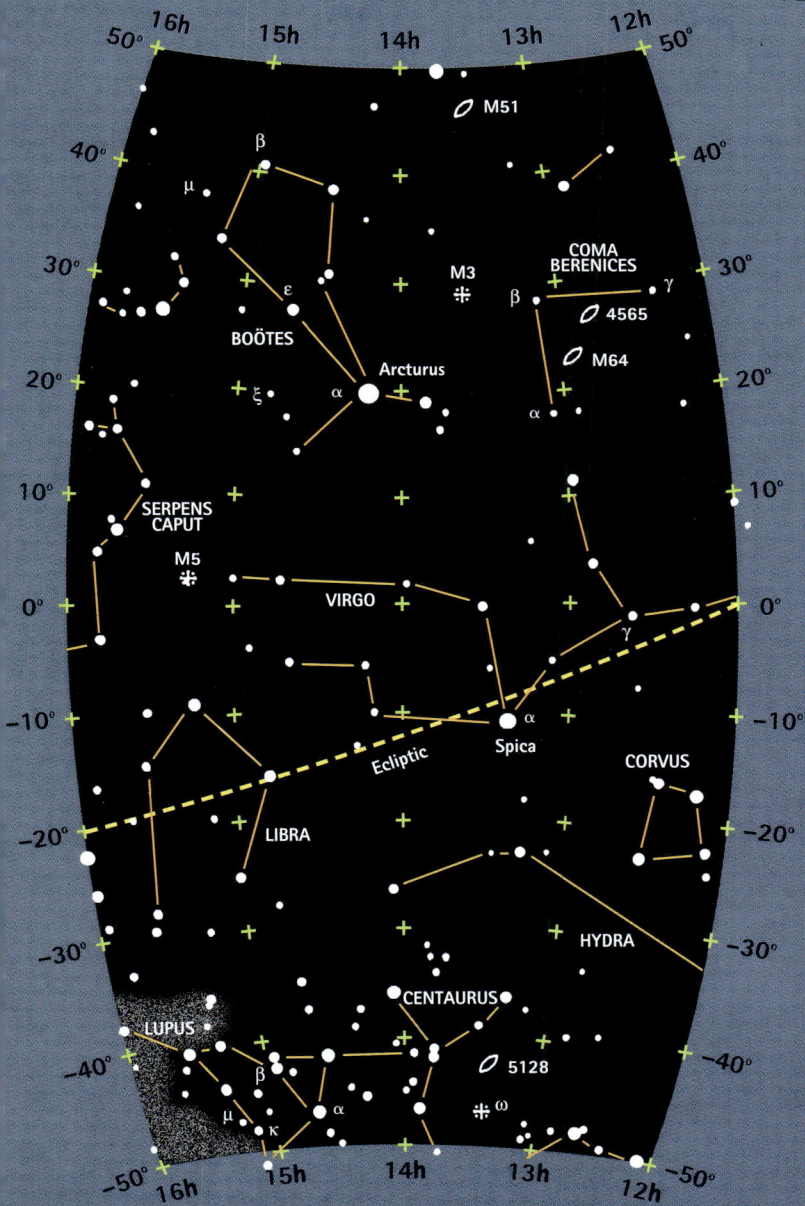

16h 15h 14h 13h 12h
50°

40°

β
μ

30° M3 COMA
BERENICES

ε β γ

4565
BOÖTES

M64
Arcturus

ξ α
SERPENS
CAPUT α

M5

VIRGO

γ

Ecliptic α
Spica CORVUS

LIBRA

HYDRA

CENTAURUS
LUPUS 5128

β ω
μ α
κ

16h 15h 14h 13h 12h

CENTAURUS, THE CENTAUR

This fine far southern constellation is one of the featured Key Constellations (see page 20).

COMA BERENICES, BERENICE'S HAIR

This is a relatively modern (1600s) constellation formed out of faint stars the Greeks considered as part of Leo's tail. It represents the long tresses the Egyptian Queen Berenice cut off to thank the gods for returning her husband, King Ptolemy, safely from battle.

The only interest to the naked eye is the loose cluster of faint stars around Gamma (γ), which look much better through binoculars. Though disappointing to the naked eye, Coma Berenices is a delight for telescope observers, as it is a region that is rich in galaxies. They lie to the south of Gamma and extend into the neighboring constellation of Virgo. They belong to the massive grouping of some 3,000 galaxies known as the Coma-Virgo cluster.

▲ Part of the spiral galaxy M100, in Coma Berenices, with its center highlighted.

The center of M100, showing clearly defined spiral arms. ▲
It is one of thousands in the Coma cluster.

The majority of these galaxies typically lie more than 400 million light-years away and so are beyond the reach of most amateur telescopes, although larger ones may just glimpse them as faint fuzzy balls. There are, however, one or two closer galaxies within the reach of small instruments. They can be found on a line running between Gamma and Alpha (α), and include M64 and NGC4565. Larger telescopes show that M64 has a large dark spot in its center, which has earned it the name of the Black Eye Galaxy. They show NGC4565 sideways as a thin line, which is why it is often called the Needle Galaxy.

LUPUS, THE WOLF

This constellation is located on the edge of the southern Milky Way in close proximity to the conspicuous constellation Centaurus. The Greeks and Romans thought of Lupus as being any old wild beast, often visualized impaled on a pole held by the centaur.

Small telescopes reveal that many of the brightest stars in Lupus are doubles. They include Kappa (κ), and Mu (μ). Mu is an easy double, and larger instruments will show that the brighter of the pair is itself a double.

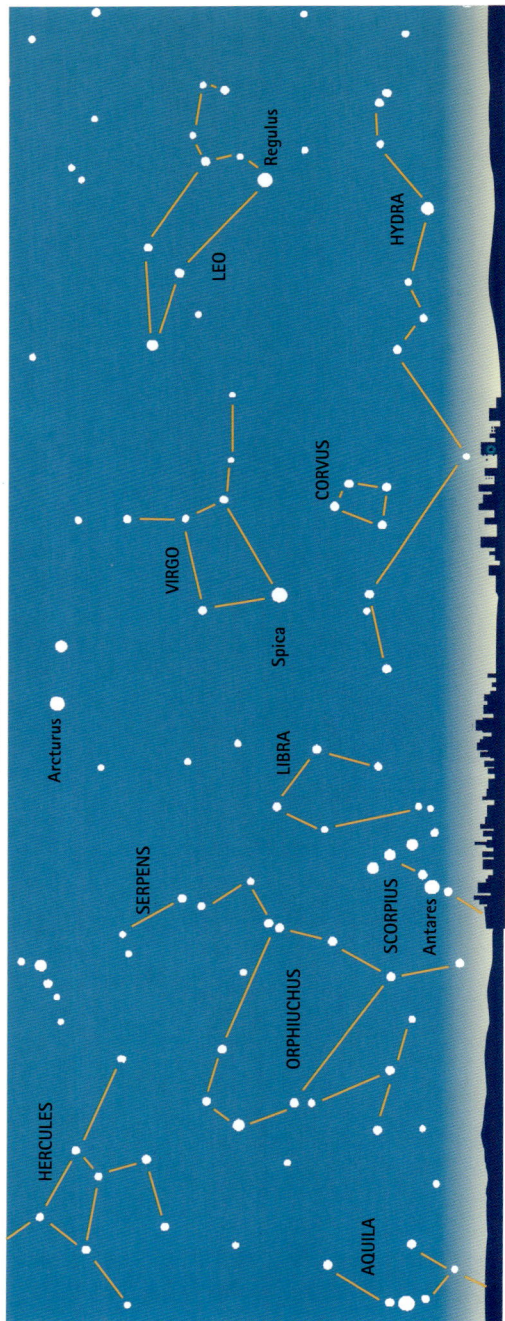

NORTHERN HEMISPHERE

EAST

SOUTH

WEST

MAY SKIES—LOOKING SOUTH
Constellations visible in North America and Europe at about 11:00 P.M. on about May 7.

Meridian skies are graced by Arcturus high up and Spica lower down. They contrast noticeably in color, with Spica being brilliant white, and Arcturus a distinctive orange. Low down on the southeast horizon the scorpion is rising, with red

Antares, marking the scorpion's heart, just visible in dark skies, giving northern observers a glimpse of southern delights. In the far east another beacon star puts in an appearance, Altair in Aquila.

NORTHERN HEMISPHERE

WEST NORTH EAST

MAY SKIES—LOOKING NORTH

Constellations visible in North America and Europe at about 11:00 P.M. on about May 7.

Cassiopeia is still low in the sky, just east of the meridian. A second bird, an eagle (Aquila) has risen above the eastern horizon to join the flying swan (Cygnus). Their two first-magnitude stars, Deneb and Altair, form with Vega in Lyra the conspicuous

Summer Triangle. Of these three, Vega appears brightest, with Deneb the least bright (as it is farther away). In absolute terms, however, Deneb is by far the most brilliant, in fact being one of the brightest stars we know in our galaxy.

SOUTHERN HEMISPHERE

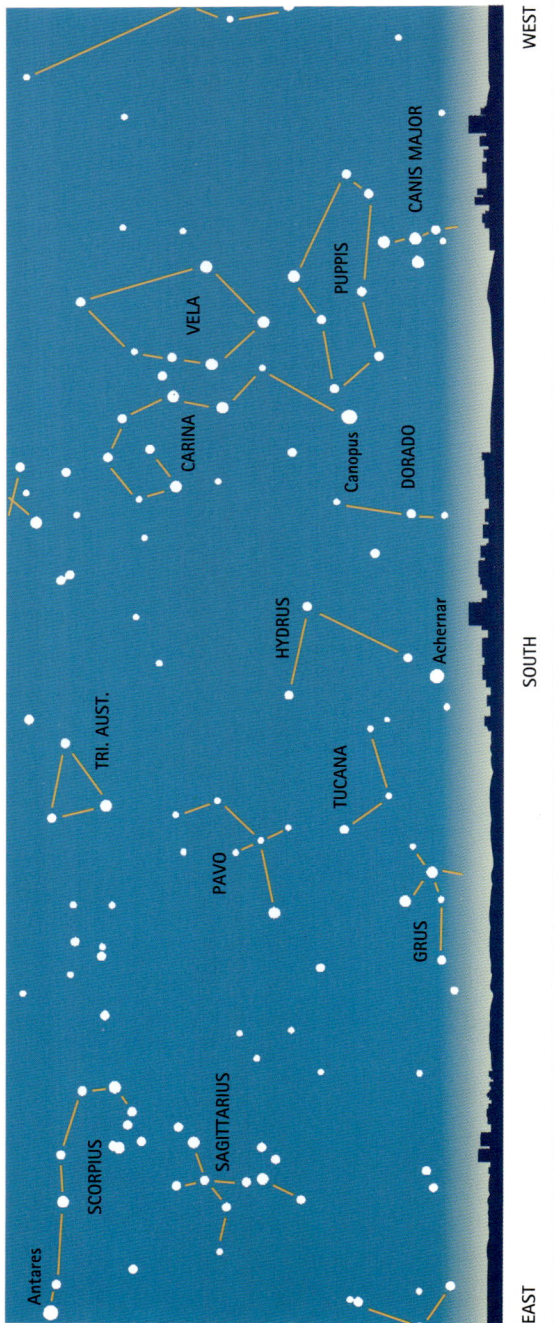

EAST

SOUTH

WEST

MAY SKIES—LOOKING SOUTH

Constellations visible in Australia and South Africa at about 11:00 P.M. on about May 7.

This month the Milky Way is particularly spectacular since it arches right across the sky. It stretches from Sagittarius, now fully risen in the east, through Canis Major, now setting in the west. Sirius has disappeared, leaving Canopus no rival,

save Achernar, which is noticeably fainter. Canopus's constellation, Carina, occupies southwest skies with Vela and Puppis. In ancient times, all three constellations formed the single constellation Argo Navis.

SOUTHERN HEMISPHERE

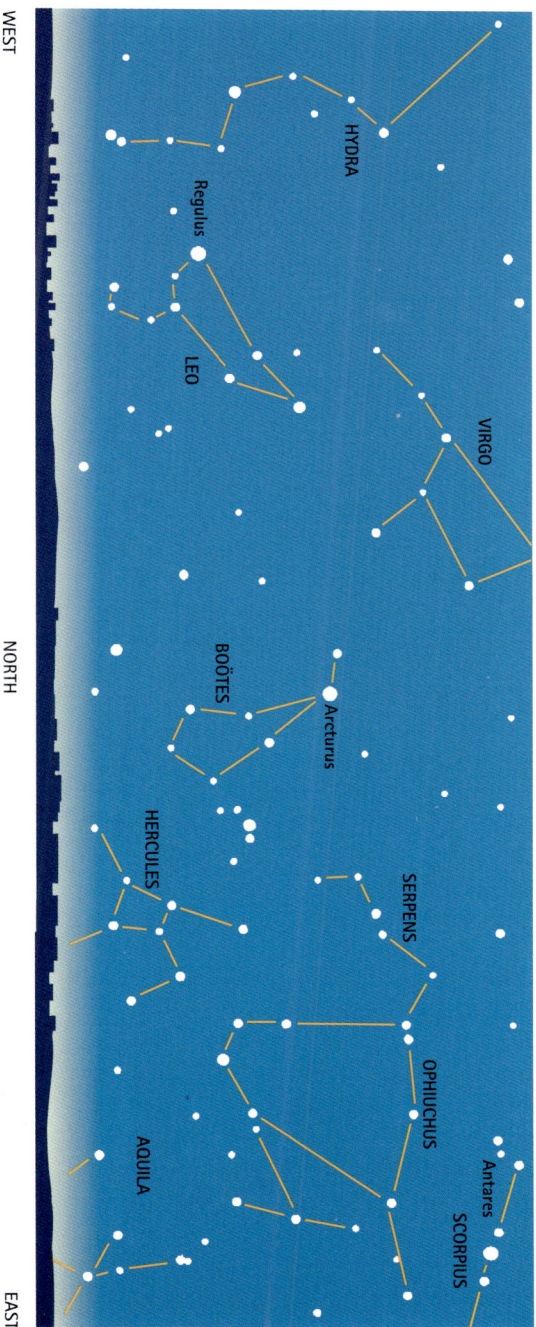

WEST

NORTH

EAST

HYDRA

Regulus

LEO

VIRGO

BOÖTES

Arcturus

HERCULES

SERPENS

OPHIUCHUS

AQUILA

Antares

SCORPIUS

MAY SKIES—LOOKING NORTH
Constellations visible in Australia and South Africa at about 11:00 P.M. on about May 7.

Arcturus in Boötes occupies center stage in meridian skies, with Spica in Virgo higher up (just out of the frame here). They contrast noticeably in color, with Spica being brilliant white, and Arcturus a distinctive orange. High in the far east another beacon star puts in an appearance, Antares. Hercules has also risen over the horizon, but in general, the eastern sky is disappointing, occupied mainly by the faint, sprawling Ophiuchus. The western sky is little better, with Leo soon to set.

June Stars

The skies north and south brighten overall this month, with Hercules joining Boötes in the north and Scorpius prominent in the south. Its brightest star, Antares, marking the Scorpion's heart, forms a neat triangle with Arcturus in Boötes and Spica in Virgo. It is noticeably redder than the orange Arcturus.

CORONA BOREALIS, THE NORTHERN CROWN

This is the more prominent of the two crowns depicted in the heavens. (The other, Corona Australis, or the Southern Crown, lies in the Southern Hemisphere at the feet of Sagittarius, but none of its stars are brighter than the fourth magnitude.) Corona Borealis is the crown of the princess Ariadne, daughter of King Minos of Crete. When she married Dionysus, he flung the crown into the heavens, whereupon its sparkling jewels were transformed into stars.

The constellation's brightest star, Alpha (α), appropriately named Gemma, meaning jewel, is of the second magnitude. Making a triangle with Delta (δ) and Gamma (γ) is R. This is a highly unusual variable star, which stays at about the sixth magnitude for most of the time and is thus easily seen with binoculars. However, it may suddenly fade within a few weeks to less than the tenth magnitude and disappear from binocular view. It may remain dim for a few weeks or sometimes several months. Astronomers think that this happens because the star periodically blasts off clouds of sooty matter, which blots out its light from us.

HERCULES

Hercules is a sprawling northern constellation, covering the fifth-largest area of the sky. It is named for the Greek hero Hercules (or Heracles), who, according to legend, is the strongest man who ever lived. He was renowned for having accomplished 12 seemingly impossible tasks, or labors, to atone for having killed his wife and children while under an evil spell. Among these labors, he strangled a lion (associated with the constellation Leo), beheaded a multiheaded monster (the Hydra), and overcame man-eating horses.

The brightest star, Alpha (α), is faintly orange in hue, and is a red giant variable. But the brightness change, between the third and fourth magnitudes, is not easy to follow and takes place at no set interval.

JUNE SKIES ▶
Constellations visible near the meridian at about
11:00 P.M. during the first week in June.

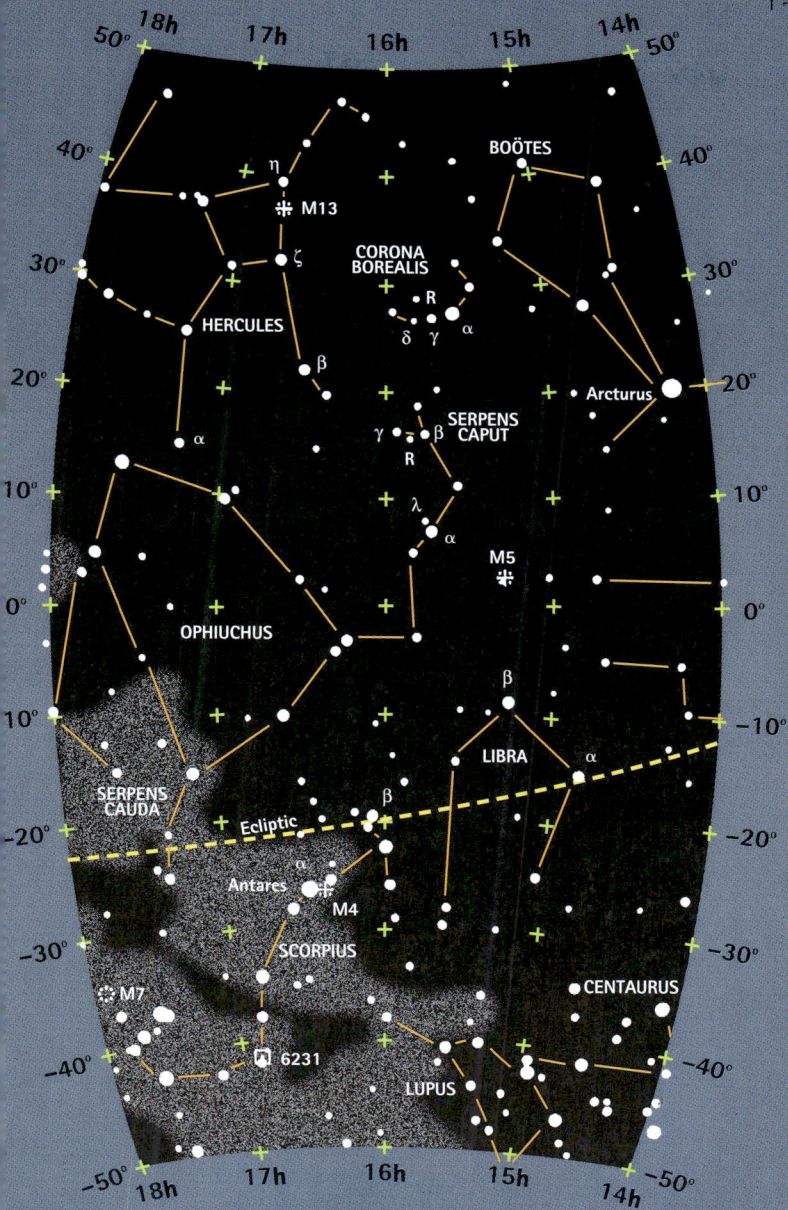

▲ The magnificent globular cluster M13 in Hercules.

A small telescope reveals that it is a double star, the fainter component being greenish in color.

However, the finest object in Hercules is M13, the finest globular cluster in northern skies. It ranks in spectacle with Omega Centauri and 47 Tucanae in far southern skies. With a magnitude of about six, it is just visible to the naked eye and easily picked up with binoculars. Telescopes reveal it as an enormous congregation of stars, clustering closely together; astronomers figure that it contains hundreds of thousands of stars altogether.

LIBRA, THE SCALES

Libra is one of the faintest constellations of the zodiac. The Sun passes through Libra between October 31 and November 23 every year.

Libra is usually associated with the scales of justice, held by the adjacent figure of Virgo. Alpha (α) is a double star, easily separated when viewed through binoculars. Beta (β) is actually slightly brighter than Alpha and has an unusual greenish tinge. Both have Arab names that sound like Star Wars characters: Alpha is named Zubenelgenubi and Beta, Zubeneschamali. Respectively, the names mean Southern Claw and Northern Claw, reflecting the fact that they were once included in the adjacent constellation Scorpius.

SCORPIUS, THE SCORPION

This constellation of the zodiac is featured as the Key Constellation this month (see page 68).

SERPENS CAPUT, THE SERPENT'S HEAD

Serpens is the only one of the 88 constellations that is split. The large Ophiuchus (Serpent Bearer) divides Serpens into Caput, the head, and Cauda, the tail. The four main head stars look nice through binoculars, which may also spot R, between Beta (β) and Gamma (γ). R is a Mira variable that brightens to the sixth magnitude at maximum but fades beyond binocular range at minimum. The constellation also boasts M5, one of the finest globular clusters in northern skies, right on the limit of naked-eye visibility. It is best found by extending south a line through Lambda (λ) and Alpha (α).

Many white dwarfs (circled) have been found in the M4 globular cluster in Scorpius. ▼

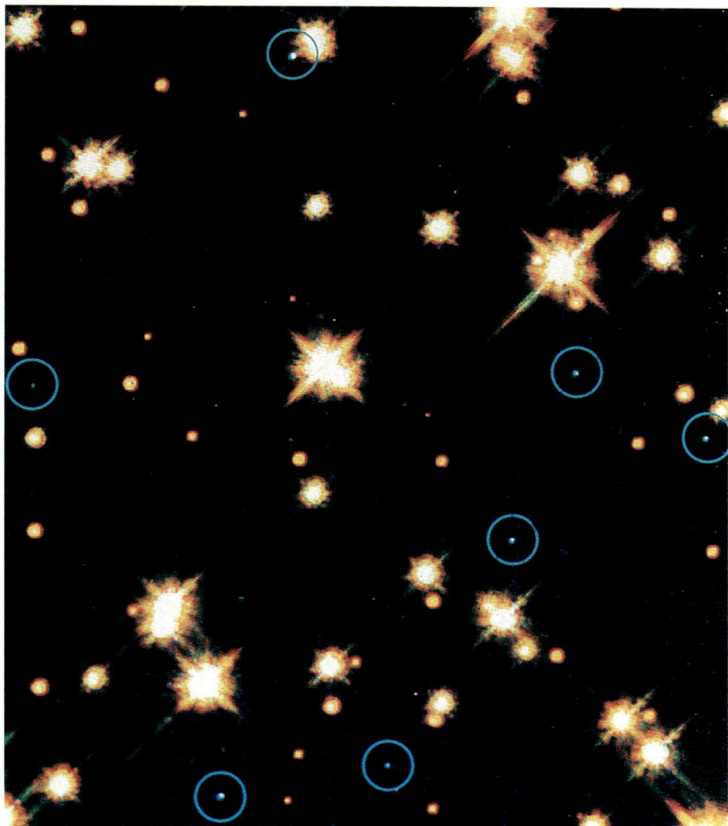

Scorpius, THE SCORPION

This is one of the few constellations that actually look like the figures they are supposed to represent. Its bright stars outline the spread claws, body, and wickedly curved tail of a scorpion poised ready to strike with its deadly sting. In Greek mythology it was none other than the mighty hunter Orion, whom the Scorpion stung to death. In the heavens, the two constellations can never be seen together—Orion sets as Scorpius rises.

Scorpius is one of the constellations of the zodiac, called Scorpio by astrologers, incidentally. The Sun passes through Scorpius between November 23 and 29 every year.

RIVAL OF MARS

Outshining all the other stars in Scorpius is Antares. Its name means "rival of Mars" because it is a distinctly red star, with a similar hue to the "Red Planet," Mars. Antares is also surrounded by a great red cloud, which shows up in long-exposure photographs. This star is a supergiant,

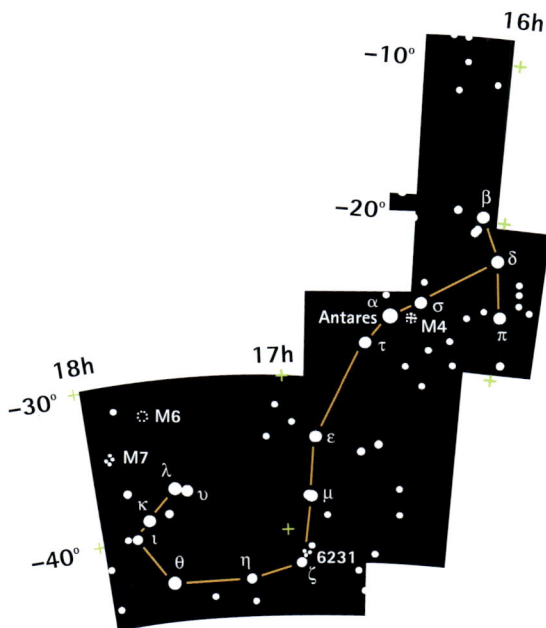

hundreds of times bigger than the Sun. Like many huge stars, it is unstable and pulsates, brightening and dimming over a period of about five years.

Scorpius is embedded in one of the richest regions of the Milky Way. Scanning the constellation with binoculars is hugely rewarding, offering a veritable feast of star clouds and clusters, and glowing nebulae. This region is so full of astronomical delights because it is close to the center of our Galaxy, which is in neighboring Sagittarius.

Many of the deep-sky objects are visible to the naked eye and stunning in binoculars. They include, near the star Zeta (ζ) in the tail, the open cluster NGC6231, formed of young hot stars. North of the tail are two other clusters, M6 and M7. Binoculars will reveal that M6 has stars arranged like an insect with open wings, which has earned it the name of the Butterfly cluster. Of about the fourth magnitude, M6 is easily seen with the naked eye. So is the third magnitude, M7, a broader cluster standing out against the Milky Way.

Brightest of the fine globular clusters in the constellation is M4, close to Antares. It is just on the limit of naked-eye visibility but can be difficult to make out because of the glare of Antares. Binoculars will find it immediately because it is in the same field of view as Antares.

NORTHERN HEMISPHERE

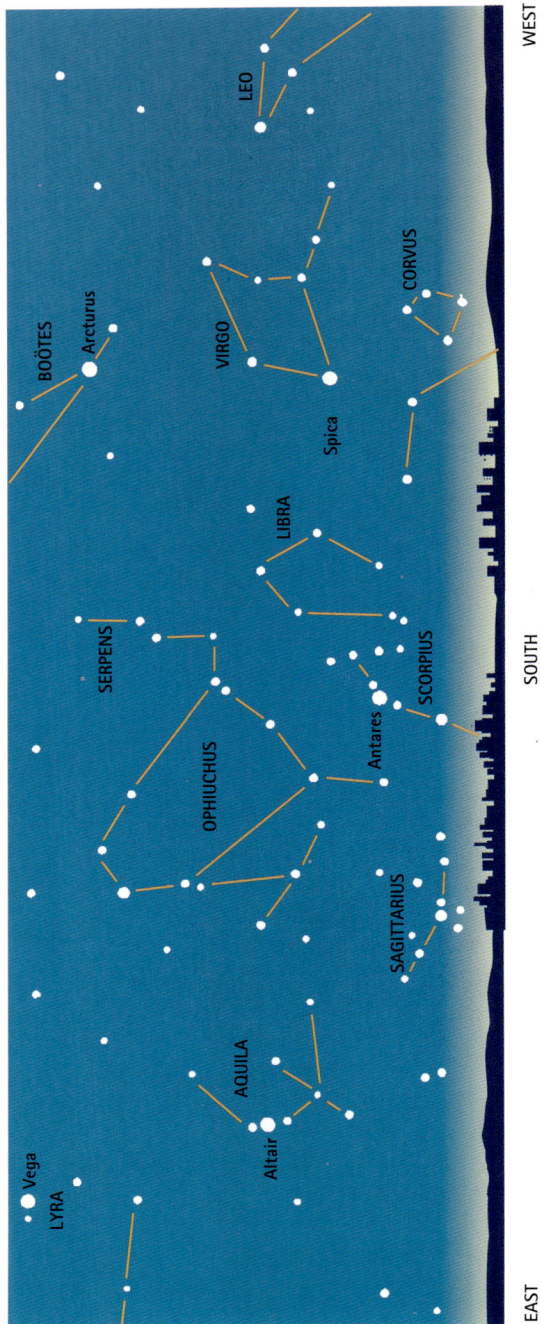

WEST

LEO

BOÖTES

Arcturus

CORVUS

VIRGO

Spica

SOUTH

LIBRA

SERPENS

SCORPIUS

OPHIUCHUS

Antares

SAGITTARIUS

AQUILA

Altair

Vega

LYRA

EAST

JUNE SKIES—LOOKING SOUTH

Constellations visible in North America and Europe at about 11:00 P.M. on about June 7.

With faint Ophiuchus moving toward the meridian, flanked by the serpent's head and tail, much of the sky is bland. But ringing this relatively bare area is a broad curve of bright stars, beginning with Vega high in the east, followed going west by Altair,

Antares, Spica, and, high above Spica, Arcturus. Leo now is close to setting beneath the western horizon. But another southern delight, Sagittarius, has just risen in the southeast.

NORTHERN HEMISPHERE

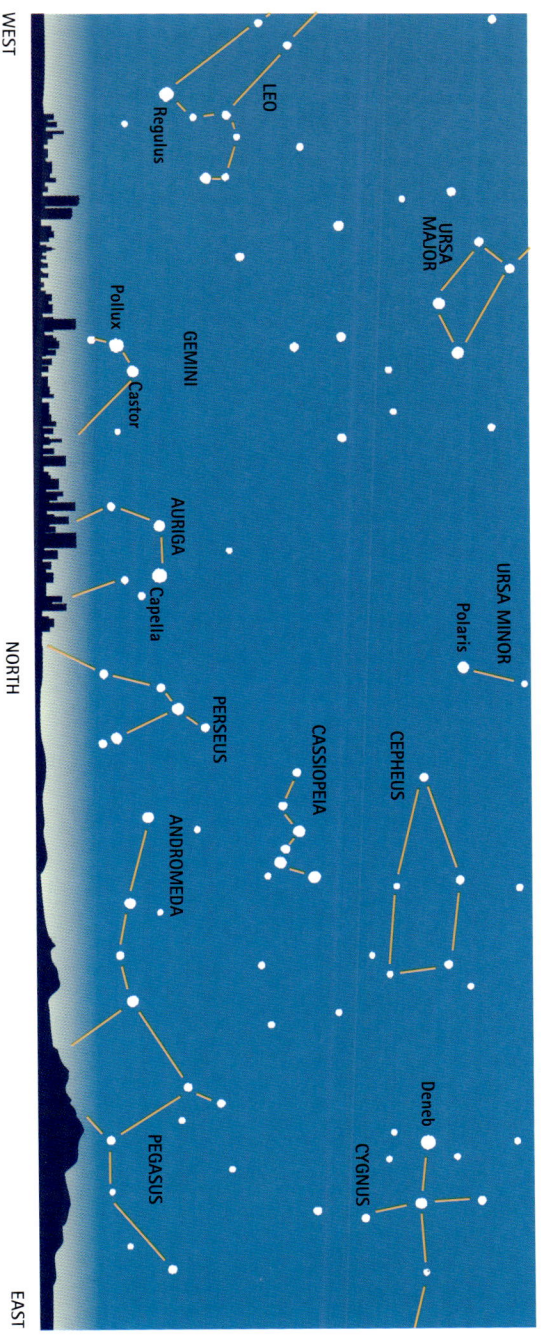

WEST

NORTH

EAST

JUNE SKIES—LOOKING NORTH

Constellations visible in North America and Europe at about 11:00 P.M. on about June 7.

The Big Dipper comes into our view in the west, while Cassiopeia continues to climb in the east. Deneb and the other two stars of the Summer Triangle, Altair and Vega, are climbing high too. Meanwhile, Andromeda and Pegasus have appeared over the eastern horizon. But the faint misty patch of the Andromeda Galaxy may be difficult to make out in the relatively light summer skies. However, observers should have no trouble spotting the four bright stars strung roughly in line in the west—Regulus, Pollux, Castor, and Capella.

Labels on chart: LEO, Regulus, URSA MAJOR, URSA MINOR, Polaris, Pollux, Castor, GEMINI, AURIGA, Capella, PERSEUS, CEPHEUS, CASSIOPEIA, ANDROMEDA, Deneb, CYGNUS, PEGASUS

SOUTHERN HEMISPHERE

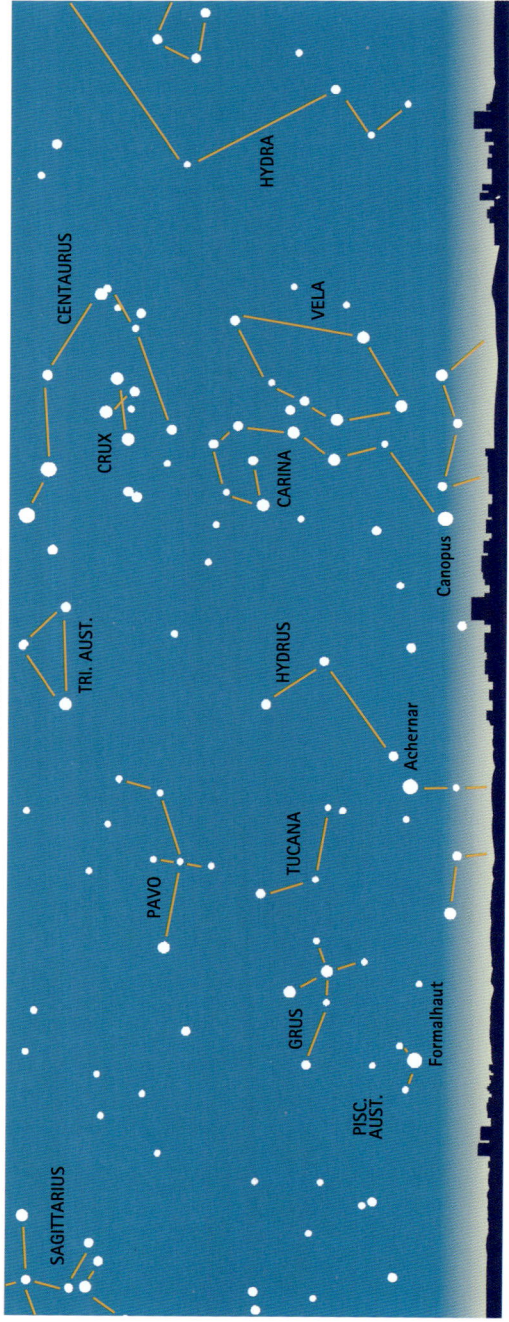

EAST

SOUTH

WEST

JUNE SKIES—LOOKING SOUTH

Constellations visible in Australia and South Africa at about 11:00 P.M. on about June 7.

The aptly named southern triangle (Triangulum Australe) sits on the meridian this month. The brightest part of the sky ahead is in the southwest, where Centaurus and Crux have reappeared. Carina and Vela are still evident lower down, with Canopus close to the horizon. The skies of the southeast are dominated by three of the southern birds—the peacock (Pavo), toucan (Tucana), and crane (Grus). The southern fish (Piscis Austrinus), with lead star Fomalhaut, is just rising—perhaps unwisely—beneath the crane.

SOUTHERN HEMISPHERE

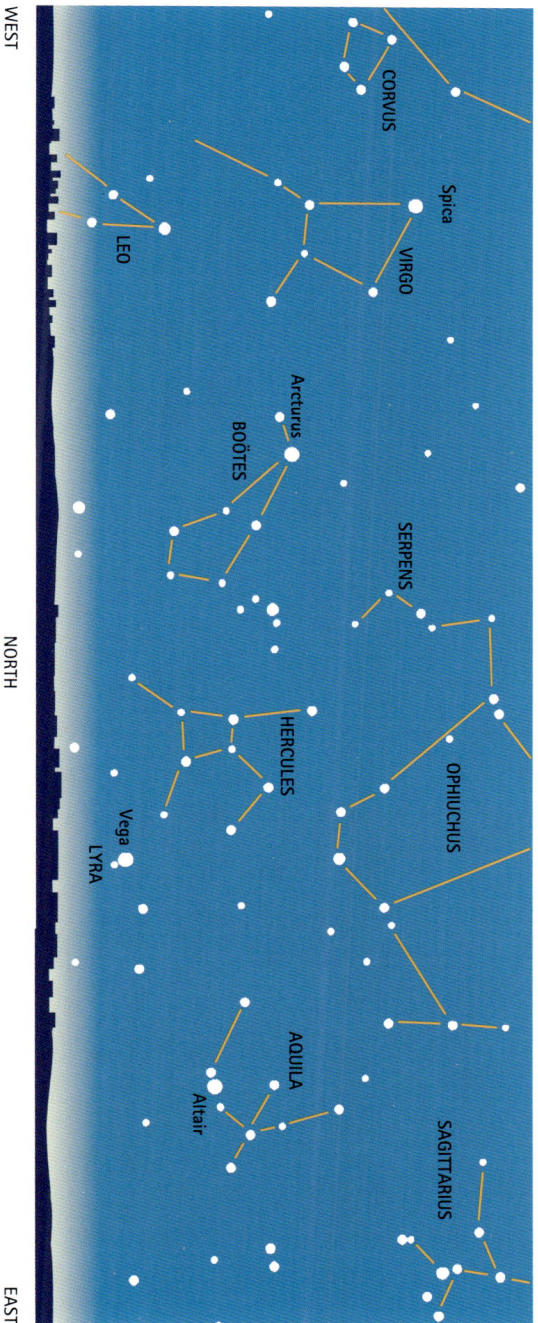

WEST

NORTH

EAST

JUNE SKIES—LOOKING NORTH

Constellations visible in Australia and South Africa at about 11:00 P.M. on about June 7.

With faint Ophiuchus moving toward the meridian, flanked by the serpent's head and tail, much of the sky is bland. But ringing this relatively bare area is a broad curve of bright stars, beginning with Spica high in the west, followed going east by Arcturus,

Vega (close to the horizon), and Altair. The familiar northern constellations Boötes and Hercules appear on either side of the meridian, with the curve of Corona Borealis between them. Only the tail of Leo is visible above the western horizon.

July Stars

This month sees the return to northern skies of the three stellar beacons that herald summer. These magnificent three—Deneb in Cygnus, Vega in Lyra, and Altair in Aquila—form the conspicuous Summer Triangle. The name also tends to be used by observers in the Southern Hemisphere, although there, of course, it is now winter! The southern skies this month are dazzling, with Scorpius still prominent and Sagittarius in hot pursuit as the heavens revolve. Northern observers, alas, can catch only a glimpse of these magnificent constellations.

LYRA, THE LYRE

The lyre in question was the instrument played by the most celebrated musician in Greek mythology, Orpheus. His hauntingly beautiful music, it was said, caused even the rivers to stop flowing so that they could listen.

Lyra is only a tiny constellation, but it is packed full of interest. It has only one bright star, Vega, which is the fifth-brightest star in the heavens. It is sometimes called the Harp Star, but its name means "swooping eagle" in Arabic, for the Arabs thought of the constellation as an eagle. Historically, Vega was the Pole Star in about 10,000 B.C. and will become so again in A.D. 14,500. This is the result of the Earth's axis gradually changing direction in space, a movement known as precession.

▲ One of Lyra's best-known features, the planetary nebula that looks just like a smoke ring. It is the Ring Nebula.

JULY SKIES ▶
Constellations visible near the meridian at about 11:00 P.M. during the first week in July.

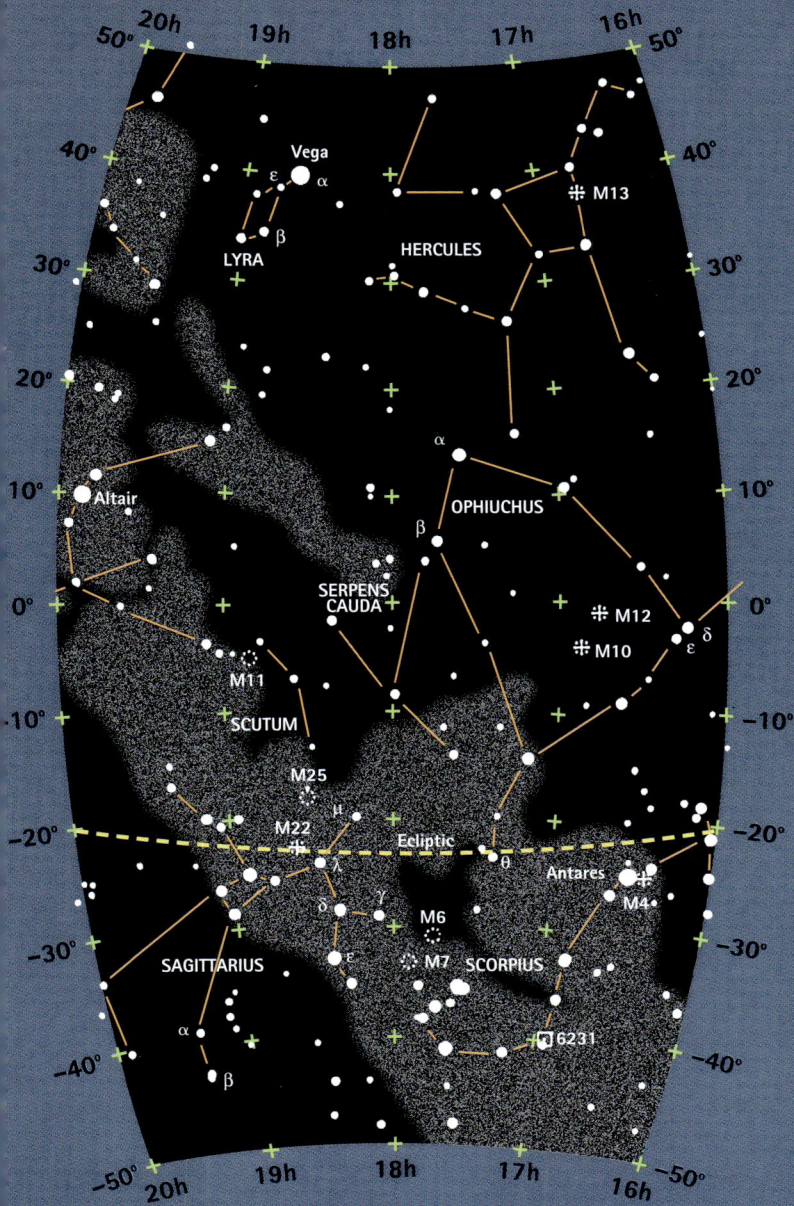

20h   19h   18h   17h   16h

50°   50°

40°   40°

Vega
ε   α
LYRA   β   HERCULES   ⊞ M13

30°   30°

20°   20°

Altair   α
OPHIUCHUS
β   ⊞ M12
SERPENS   ⊞ M10
CAUDA   ε δ

10°   10°

0°   0°

M11
SCUTUM

-10°   -10°

M25
μ
M22   Ecliptic
θ
Antares
M4
λ   Ⓧ
δ   γ
M6
SAGITTARIUS   ε   ⊞ M7   SCORPIUS

-20°   -20°

-30°   -30°

α   ⊡ 6231

-40°   -40°

β

-50°   -50°

20h   19h   18h   17h   16h

Close to Vega is a star that is a particular favorite among northern astronomers. It is the "double-double" Epsilon (ε). The very sharp-sighted may be able to spot that it is a double star with the naked eye, and the pair of stars are easily separated through binoculars. Seen through a small telescope, each star of the pair can be seen to be a double too. Remarkably, each of the four stars has about the same brightness, which is unusual among doubles, let alone double-doubles.

Beta (β) is another fascinating star. It is a variable of the eclipsing binary type, made up of a small bright star and a large dimmer star. The two revolve around each other, and periodically (every 13 days) the dim one covers the bright one, causing Beta to fade temporarily.

OPHIUCHUS, THE SERPENT BEARER

This is a large constellation depicting a man carrying a snake. Since the snake sheds its skin every year, it has long been a symbol for renovation and healing. Therefore, the Greeks associated Ophiuchus with Asclepius, their god of medicine, who was capable even of reviving the dead.

As a constellation, Ophiuchus splits the snake (Serpens) in two—Caput (Head) and Cauda (Tail). Its stars are not especially impressive, with Alpha (α) being only of the second magnitude. Its name is Rasalhague, Arabic for "head of the serpent catcher."

The interior of the constellation is relatively barren, but boasts two fine globular clusters (M10 and M12), found east of the pair of stars Delta (δ) and Epsilon (ε). There are more clusters to be seen in the Milky Way at the Serpent Bearer's feet. They include M19, midway between Theta (θ) and the bright Antares in the adjacent Scorpius.

SAGITTARIUS, THE ARCHER

The Sun passes through this constellation of the zodiac between December 18 and January 19 every year. It is one of the most spectacular of all the constellations, embedded in the richest region of the Milky Way. The center of our galaxy lies in this direction.

The ancient Greeks thought of the figure as Crotus, who invented archery and was the son of the pipe-playing god Pan. He is depicted as a centaur, half-man, half-horse. He is pulling a bow, with the arrow aimed at Antares, the heart of the Scorpion.

The group of five stars in the center of the constellation is often called the Milk Dipper after its shape and the fact that it "dips" into the Milky Way. Add Gamma (γ), Epsilon (ε), and Delta (δ) to the group, and it becomes the Teapot, with a pointed lid and long spout.

The individual stars in the constellation are not particularly interesting, although Beta (β) is a naked-eye and binocular double. It is in the surrounding Milky Way that Sagittarius is so spectacular. There is a host of globular clusters, none more magnificent than M22, the third-brightest globular in the heavens and visible to the naked eye. Further north is M25, a fine open cluster when viewed in binoculars. The region west of the stars Lambda (λ) and Mu (μ) is especially rich in clusters and nebulae, which include the glorious Lagoon (M8) and Trifid (M20).

SERPENS CAUDA, THE SERPENT'S TAIL

The tail end of the divided constellation Serpens is much less impressive than the head (Serpens Caput, see page 67). Its southern end dips into the Milky Way, on the border with Sagittarius. In this region, just west of Gamma (γ) in neighboring Scutum, is the delightful Eagle nebula (M16). This is the site of one of the Hubble Space Telescope's most dramatic photos, named The Pillars of Creation.

Clouds in the Eagle Nebula, where stars are being born. ▼

NORTHERN HEMISPHERE

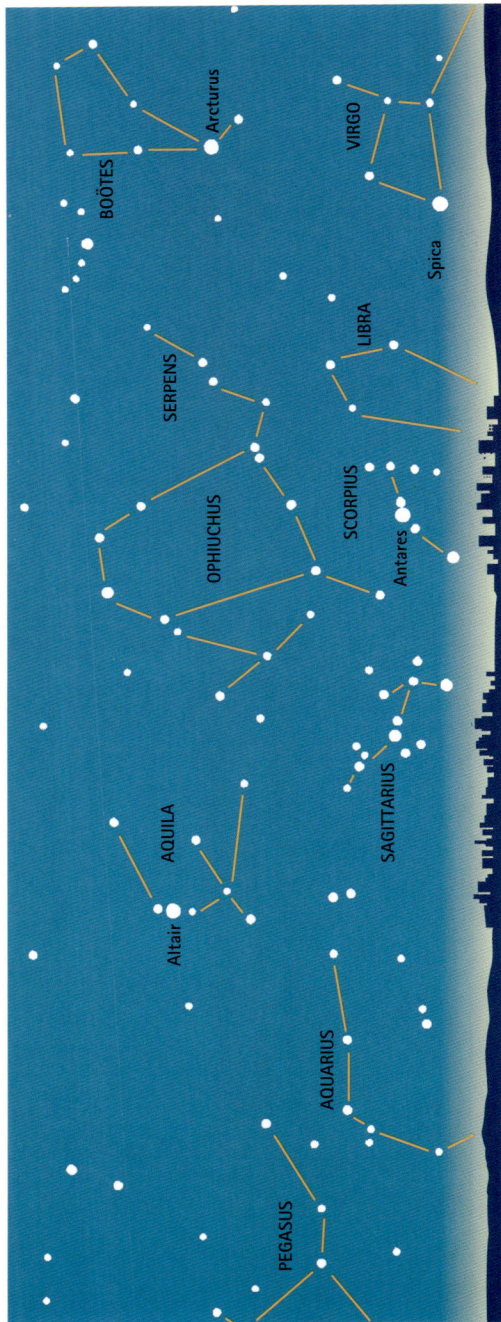

WEST

BOÖTES

Arcturus

VIRGO

Spica

SERPENS

LIBRA

OPHIUCHUS

SCORPIUS

Antares

SAGITTARIUS

AQUILA

Altair

AQUARIUS

PEGASUS

EAST

SOUTH

JULY SKIES—LOOKING SOUTH

Constellations visible in North America and Europe at about 11:00 P.M. on about July 7.

July is another good month to peer deep into the southern hemisphere, although it would be better if summer skies were darker. This is the month when observers can see Sagittarius furthest above the horizon, although Scorpius is beginning to set.

Arcturus and Spica (low down) form a bright pair in the west. Aquila is coming up to the meridian, with Altair brilliant. With Deneb in Cygnus and Vega in Lyra overhead (and out of the frame here), Altair forms the prominent Summer Triangle.

NORTHERN HEMISPHERE

WEST   NORTH   EAST

JULY SKIES—LOOKING NORTH

Constellations visible in North America and Europe at about 11:00 P.M. on about July 7.

Both bears, Ursa Major and Minor, are now descending. Regulus has disappeared beneath the western horizon and the rest of the constellation of Leo will soon be following. Capella is conspicuous, now low down on the northern horizon as Boötes

straddles the meridian. Cassiopeia still delights in mid-skies to the east, the starry background of the Milky Way as ever providing a feast in binoculars. Andromeda is better placed now for observation, and the Square of Pegasus has fully risen.

SOUTHERN HEMISPHERE

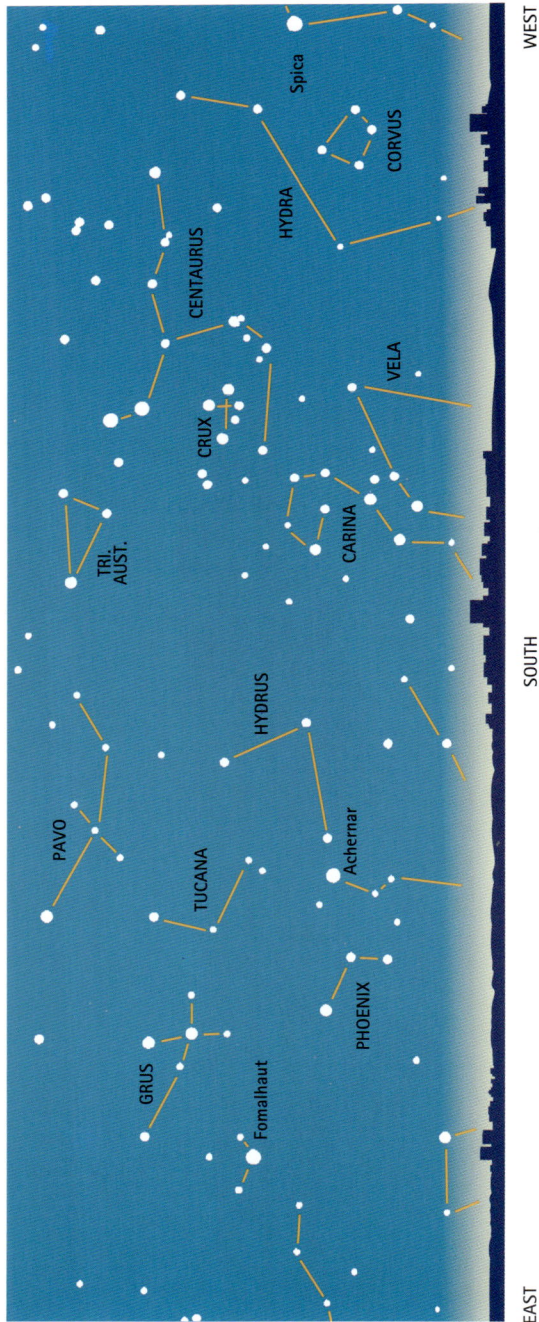

EAST

SOUTH

WEST

JULY SKIES—LOOKING SOUTH

Constellations visible in Australia and South Africa at about 11:00 P.M. on about July 7.

The dazzling Milky Way stands nearly vertical this month just west of the meridian, running from Scorpius high overhead (out of the frame here) through Carina and Vela, now sinking beneath the horizon. Canopus has disappeared, showing that it is not quite circumpolar, unlike Achernar, which never sets and is now climbing in the southeast. The only other really bright star in eastern skies is Fomalhaut, not outstanding among first-magnitude stars, but appearing so in a generally dull region of the heavens.

SOUTHERN HEMISPHERE

WEST

NORTH

EAST

JULY SKIES—LOOKING NORTH

Constellations visible in Australia and South Africa at about 11:00 P.M. on about July 7.

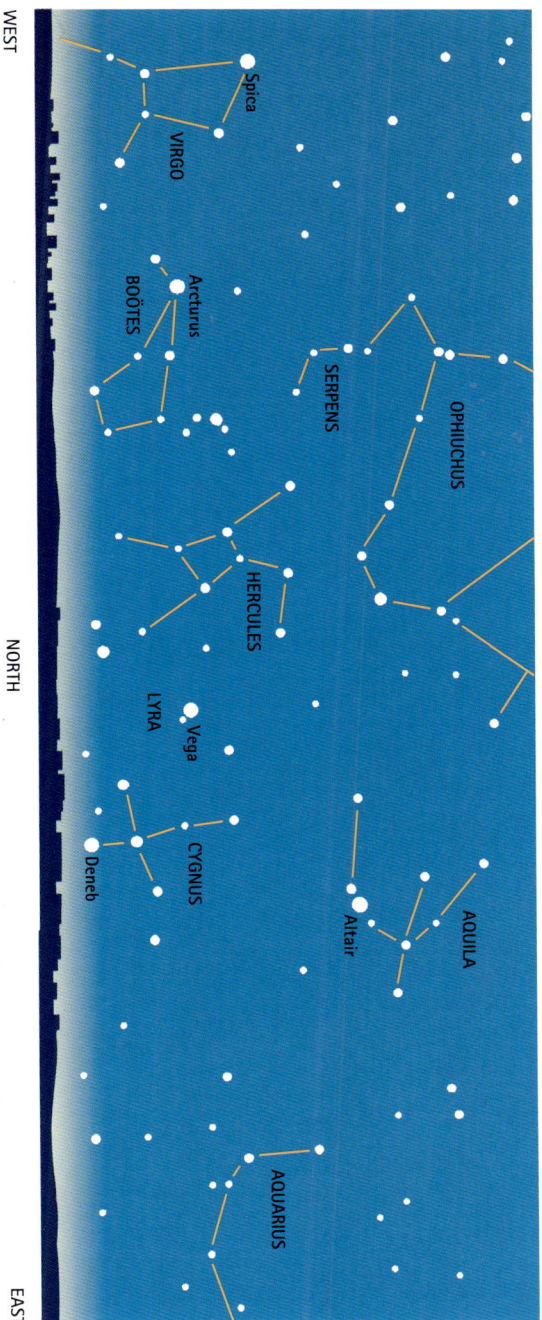

Arcturus and Spica are now low in the west and will shortly disappear, as will Hercules. Above them the skies lack interest, occupied mainly by Ophiuchus and Serpens. Vega in Lyra is close to the meridian. Deneb is just appearing over the

horizon further east, while higher up is Altair. These three stars make up the so-called Summer Triangle. This is not an appropriate name for southern observers, since, while it is summer in the northern hemisphere, it is winter in the southern.

August Stars

The Summer Triangle still blazes in northern skies, this month straddling the meridian in late evening. On moonless nights, the Milky
Way softly glows overhead, bisecting the sky. It is much fragmented in
this aspect by dark lanes of impenetrable dust.

CYGNUS, THE SWAN

The distinctive northern grouping of Cygnus is the featured constellation
this month (see page 86).

AQUILA, THE EAGLE

This second prominent bird wings its way through the Milky Way just
south of Cygnus, straddling the celestial equator. In mythology, Aquila
was the favorite bird of the king of the gods, Zeus, who used it to
retrieve the thunderbolts he hurled at his enemies.

Aquila's leading star, Altair, is the southernmost of the trio of stars
that form the Summer Triangle, along with Deneb in Cygnus and Vega in

▲ A "hole" in Cygnus is
caused by a dark cloud
of gas and dust.

AUGUST SKIES ▶
Constellations visible near the meridian at about
11:00 P.M. during the first week in August.

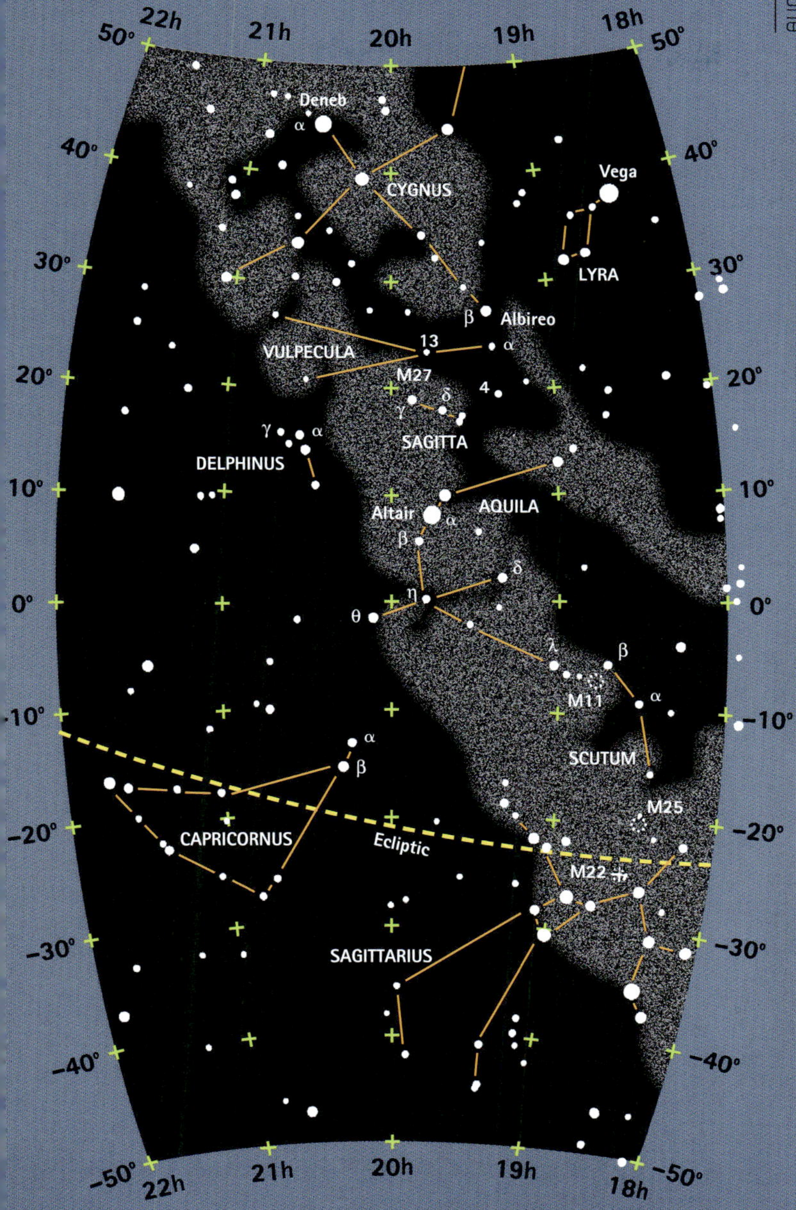
22h
21h
20h
19h
18h
50°
40°
30°
20°
10°
0°
-10°
-20°
-30°
-40°
-50°
Deneb
CYGNUS
Vega
LYRA
Albireo
13
4
VULPECULA
M27
SAGITTA
DELPHINUS
AQUILA
Altair
M11
SCUTUM
M25
M22
CAPRICORNUS
Ecliptic
SAGITTARIUS

Lyra. It is one of the closest of the bright stars, being only about 17 light-years away. Centered in the prominent straight line of stars south of Altair is Eta (η). This star is a variable, of the Cepheid type, which varies in brightness as regularly as clockwork from magnitude 3.6 to 4.5 in precisely seven days, four hours. With care, the brightness changes can be followed by comparison with Theta (θ) or Delta (δ), which are third magnitude, and Lamda (λ), which is fourth.

CAPRICORNUS, THE SEA GOAT

This small constellation of the zodiac is composed of faint stars roughly in the shape of a rather crooked triangle. The Sun passes through the constellation between January 19 and February 16. Capricornus is depicted as a fish-tailed goat, associated with the goat-headed god Pan. It is one of a number of constellations in this part of the heavens linked by a watery theme, including Aquarius (Water Bearer), Pisces (Fishes), Cetus (Whale), and Piscis Austrinius (Southern Fish).

Both Alpha (α) and Beta (β) are doubles visible with binoculars, but there is not a great deal of other interest in the constellation.

DELPHINUS, THE DOLPHIN

Delphinus is a tiny constellation, but one that needs only a little imagination to picture as a dolphin leaping gracefully through the waves. In Greek mythology, the dolphin was immortalized in the heavens for its help in bringing one of the beautiful sea nymphs, the Nereids, to be the wife of the sea god, Poseidon.

The distinctive parallelogram of four stars in Delphinus looks almost like a star cluster. Gamma (γ) is a fine double for small telescopes.

SAGITTA, THE ARROW

Another tiny constellation, Sagitta does look rather arrowlike. It has been associated in mythology with both an arrow shot by Hercules and Cupid's arrow. Being in the Milky Way, Sagitta is worth scanning with binoculars, for it has some really nice star fields. Binoculars and small telescopes will pick up a globular cluster (M71) midway between Gamma (γ) and Delta (δ).

SCUTUM, THE SHIELD

This small constellation is fairly modern (1600s) and has no mythogical connections. Its main item of interest is a rich open cluster, M11. It is visible through binoculars as a fuzzy ball, which a small telescope resolves into a V-shaped formation of stars. This shape looks rather like a flock of birds winging through the sky, which earns M11 the name of the Wild Duck Cluster.

▲ The Milky Way brightens August skies north and south. This is the Lagoon Nebula in far-south Sagittarius. .

VULPECULA, THE FOX

Vulpecula is a modern constellation (1600s) of faint stars south of Cygnus. Being in the Milky Way, it contains some fine starscapes for binocular viewers. Just south of star 13 and north of Gamma (γ) in neighboring Sagitta, is the planetary nebula, M27. Larger telescopes show off its distinctive shape that earns it the name of the Dumbbell nebula. Star 4, due south of Alpha (α), is set in a charming pattern of about 10 stars that go by the name of the Coathanger.

The Dumbbell Nebula in Vulpecula, M27, a fine example of a planetary nebula. ▼

Cygnus

This is another constellation that looks impressively like the figure it is supposed to represent. Little imagination is required to picture in the outline of stars a swan winging its way across the northern sky, as if flying south on migration to escape the severe cold of the far northern winter. Its curved wings are widespread, poised for a powerful downbeat; its long neck is outstretched, its tail fanned out. The X-like shape of the constellation gives it the alternative name of the Northern Cross, mimicking the famous Southern Cross of far southern skies.

In Greek mythology, the swan was one of the many disguises adopted by Zeus when he went about his illicit affairs with fair maidens. The result of his tryst with Leda, Queen of Sparta, was that Leda laid eggs, out of which hatched the twins Castor and Pollux (which feature in the constellation Gemini) and the legendary beauty Helen of Troy, with the "face that launched a thousand ships."

DELIGHTS GALORE

Brightest of the stars in the constellation is Deneb, in the swan's tail, which is a blue-white supergiant. It is very much further away than the other bright stars in the sky, lying at a distance of some 2–3,000 light-years. To appear so bright at such a distance it must have the energy

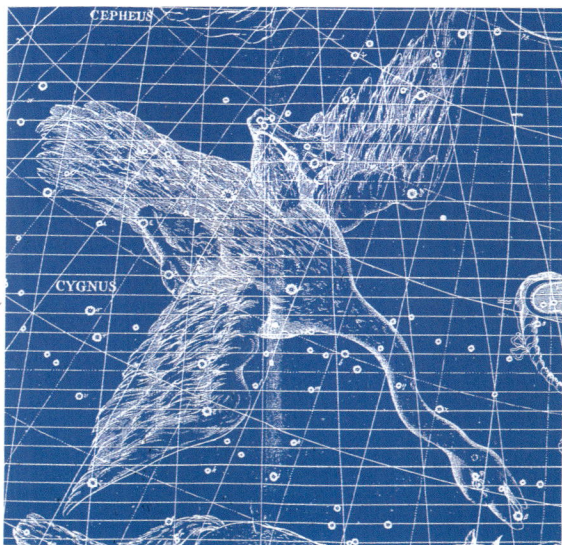

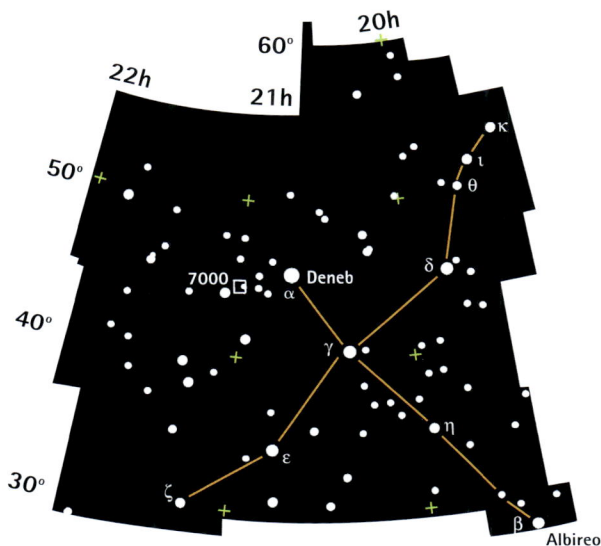

output of more than 60,000 Suns. Albireo, the star at the opposite end of the swan figure, is much fainter, but it is a fine double star, which small telescopes will separate into lovely blue and yellow components.

Cygnus is one of the richest constellations in the northern sky, as it is situated in the Milky Way. It boasts a host of nebulae and clusters. Close to Deneb, telescopes show a glowing red nebula (NGC7000) that bears an uncanny resemblance to the continent of North America, which is why it is popularly called the North American Nebula. South of Epsilon (ε), long-exposure photographs reveal a vast loop of glowing gas, in the form of wisps as delicate as lace or high cirrus clouds in Earth's skies. The Cygnus Loop is all that remains of a gigantic star that blasted itself to pieces tens of thousands of years ago.

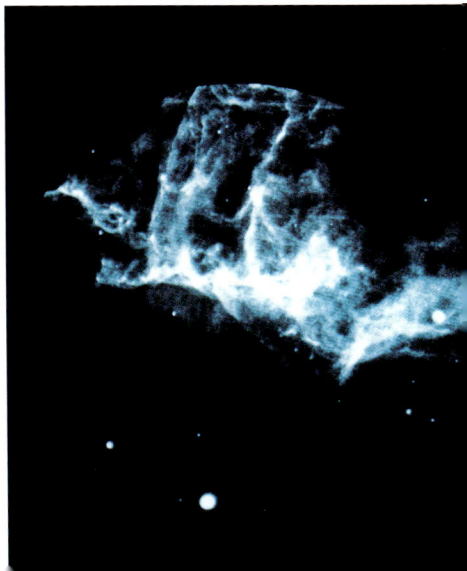

Part of the Cygnus Loop, ▶
a still-expanding cloud from an
ancient supernova explosion.

NORTHERN HEMISPHERE

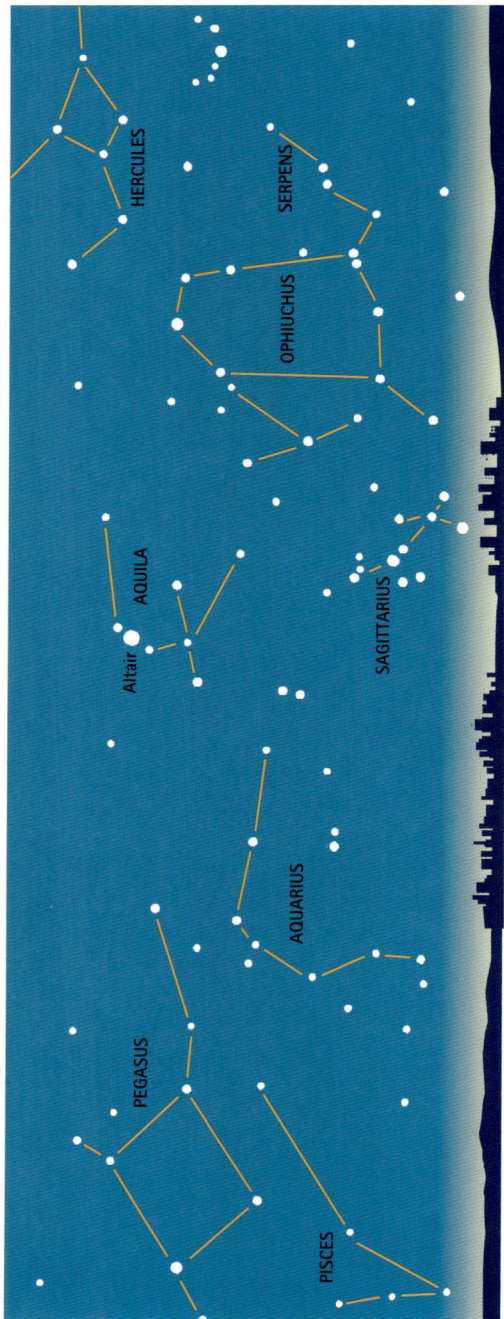

EAST

WEST

SOUTH

HERCULES

SERPENS

OPHIUCHUS

AQUILA

Altair

SAGITTARIUS

AQUARIUS

PEGASUS

PISCES

AUGUST SKIES—LOOKING SOUTH

Constellations visible in North America and Europe at about 11:00 P.M. on about August 7.

Aquila now straddles the meridian, with Altair nearly due south. The Summer Triangle it makes with Deneb and Vega (out of the frame) is almost directly overhead. This is the last month to enjoy the brilliance of Sagittarius before it slips below the southern horizon. The relatively uninteresting skies occupied by Ophiuchus and Serpens are thankfully on the way out west, not that the upcoming eastern constellations, such as Pisces and Aquarius, are particularly brilliant.

NORTHERN HEMISPHERE

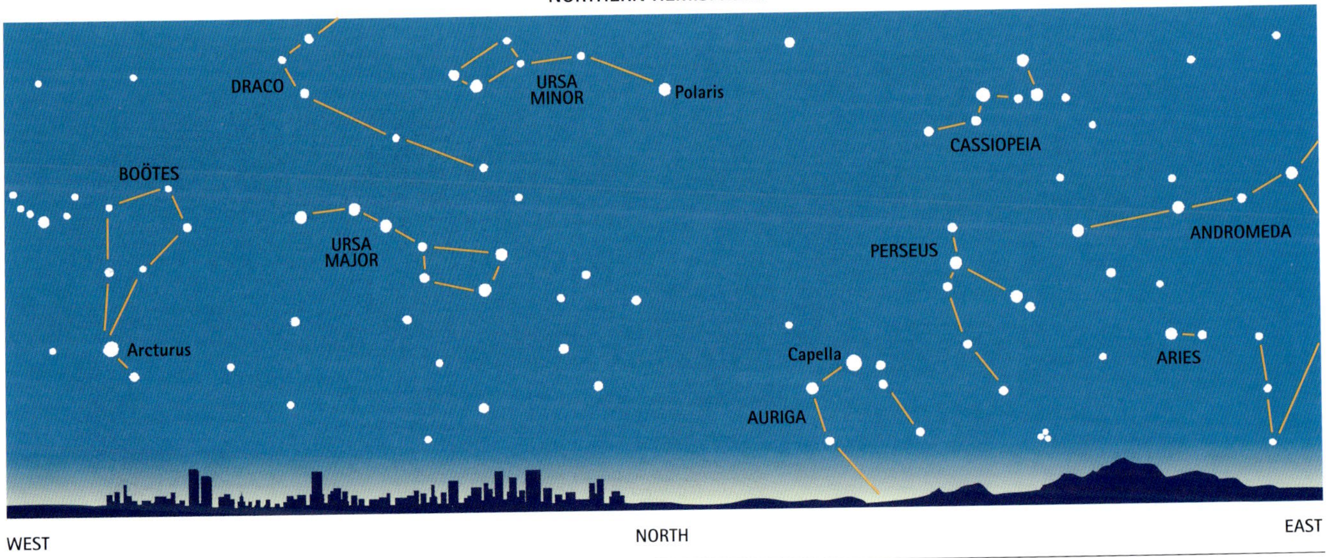
DRACO
URSA MINOR
Polaris
CASSIOPEIA
BOÖTES
URSA MAJOR
PERSEUS
ANDROMEDA
Arcturus
Capella
AURIGA
ARIES
WEST
NORTH
EAST

AUGUST SKIES—LOOKING NORTH
Constellations visible in North America and Europe at about 11:00 P.M. on about August 7.

Arcturus, which we last spotted in eastern skies in February, now makes an appearance in the west as the heavens revolve. Arcturus, at the tail end of the kite-shaped Boötes, is truly a beacon star, the fourth brightest in the whole heavens. Its appearance this month is welcome, because in this view the only other bright star is Capella, now climbing. High overhead, however, the Summer Triangle is still with us.

SOUTHERN HEMISPHERE

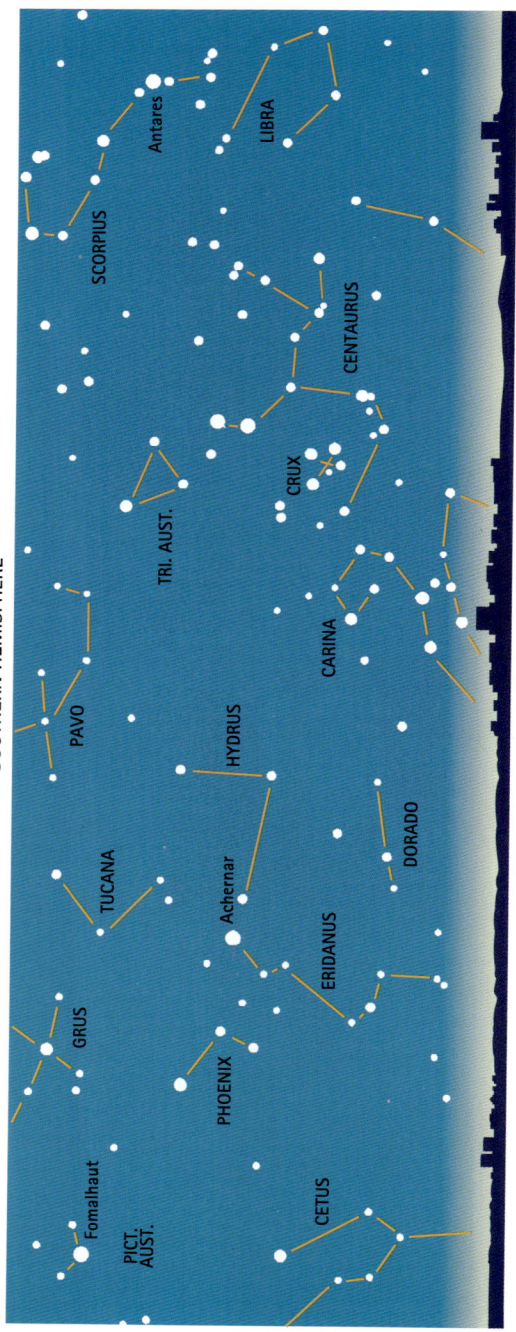

EAST

SOUTH

WEST

AUGUST SKIES—LOOKING SOUTH

Constellations visible in Australia and South Africa at about 11:00 P.M. on about August 7.

This month all the spectacle in the southern heavens is concentrated in the southwest around the Milky Way, where Scorpius can be seen again in all its glory. But the rest of the sky is dull by contrast, being occupied by a veritable flock of southern birds—the phoenix, toucan (Tucana), crane (Grus), and peacock (Pavo), which is now on the meridian. For the second month, only Achernar and Fomalhaut brighten up the east. The constellations rising above the horizon here, Cetus and Eridanus, add little interest.

SOUTHERN HEMISPHERE

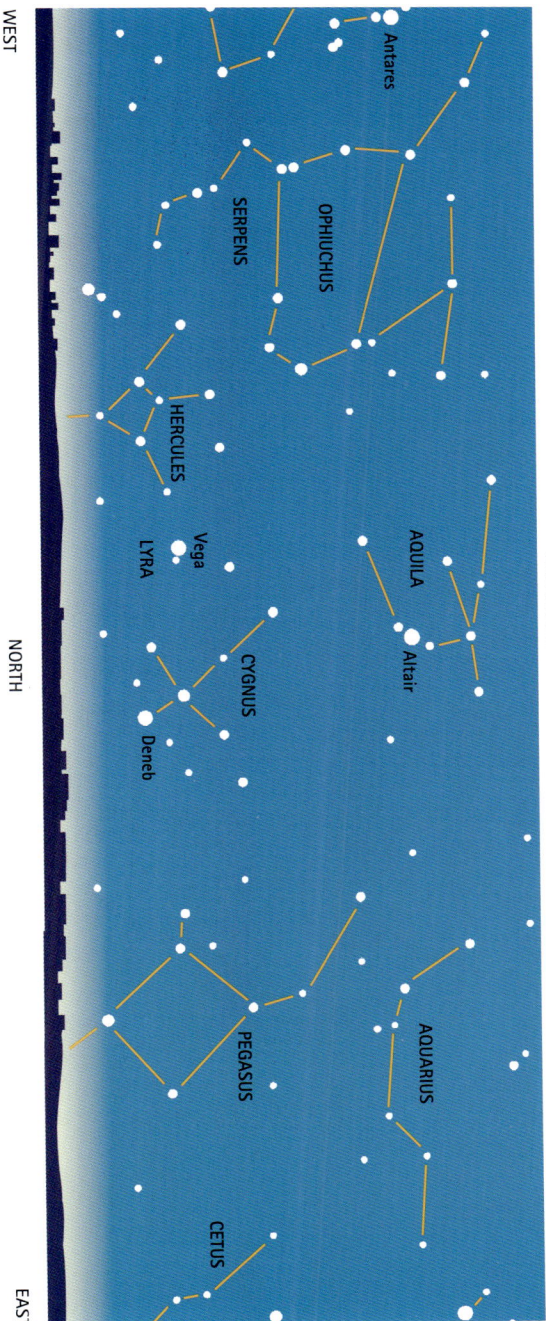

WEST

NORTH

EAST

AUGUST SKIES—LOOKING NORTH
Constellations visible in Australia and South Africa at about 11:00 P.M. on about August 7.

The Summer Triangle sits on the meridian this month, with Vega and Deneb low down and Vega in mid-skies. The only other really bright stars on show are Antares in the far west and Fomalhaut in the far east (out of the frame). Low in the east

Pegasus has risen, its famous square empty of bright stars. It is sometimes called the autumn (fall) square, but this again is relevant only to northern observers, for in southern skies Pegasus brings the promise of spring, not fall.

September Stars

The Milky Way and its necklace of glittering constellations—Cygnus, Aquila, and Sagittarius—is disappearing fast as September advances. The skies become disappointingly bland again except for Pegasus. The appearance of this constellation's famous Square signals the approach of the fall in the Northern Hemisphere and of spring in the Southern.

AQUARIUS, THE WATER BEARER

The Sun passes through this constellation of the zodiac between February 16 and March 11. It is a large constellation, though not particularly easy to make out. Aquarius depicts the figure of a youth pouring water from a jar. It pours out as a cascade of stars and ends up in the mouth of Piscis Austrinus, the Southern Fish. The figure is usually identified with the beautiful boy Ganymede, whom Zeus took back to Mount Olympus to dispense wine to the gods.

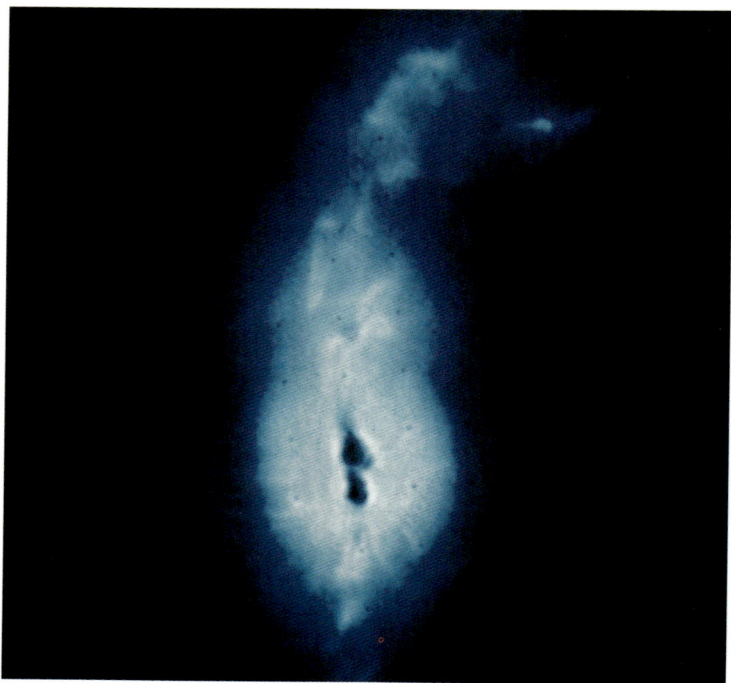

▲ Strange gas clouds surround the variable star R Aquarii.

SEPTEMBER SKIES ►
Constellations visible near the meridian at about 11:00 P.M. during the first week in September.

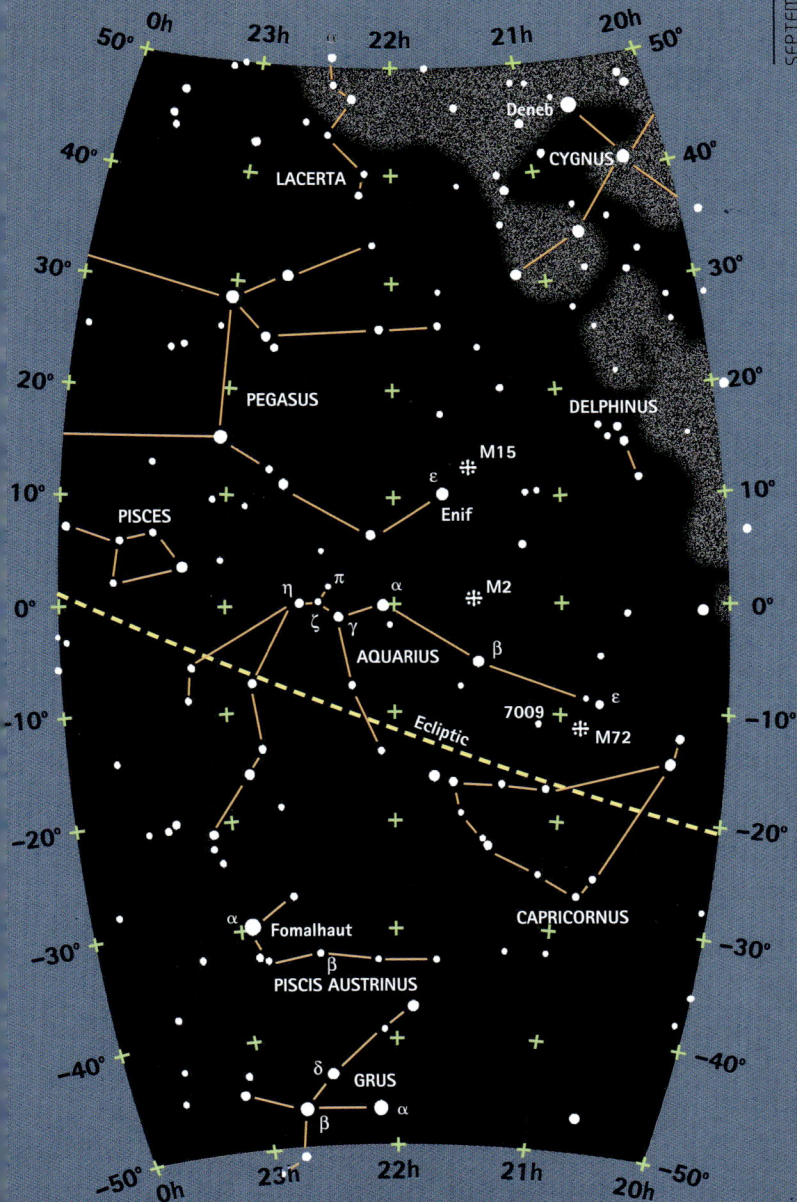

50° 0h   23h   α 22h   21h   20h   50°
40°   40°
        LACERTA   Deneb
                            CYGNUS
30°   30°
           PEGASUS
20°   DELPHINUS   20°
                                    M15
10°   ε   10°
                            Enif
     PISCES   π   α
 0°   η   ζ γ   ※ M2   0°
            AQUARIUS   β
                                    ε
-10°   Ecliptic   7009   -10°
                            ※ M72
-20°   -20°
                           CAPRICORNUS
        α Fomalhaut
-30°   β   -30°
        PISCIS AUSTRINUS
-40°   δ   GRUS   -40°
          β   α
-50° 0h   23h   22h   21h   20h   -50°

The mouth of the jar, out of which the water pours, is marked by the four stars Eta (η), Pi (π), Gamma (γ), and Zeta (ζ). Zeta is a double star, visible in telescopes. Few of the other stars are particularly interesting, but the constellation does present a few clusters and nebulae of note. They include the globular cluster M2, due west of Alpha (α) and due north of Beta (β). It is just too faint to be visible to the naked eye but is easily seen in binoculars and a small telescope.

The nebula NGC7009 can be spotted in small telescopes just south of Epsilon (ε). It is a planetary nebula, so called because it shows up as a disk, rather like a planet. Moreover, this particular planetary nebula looks somewhat like a ringed planet, which is why it is called the Saturn Nebula. Close by, and in the same field of view in binoculars, is another globular cluster, M72.

GRUS, THE CRANE

Grus is one of the better defined of the southern birds. It is a relatively modern (1600s) constellation that was, for many years, referred to as the Flamingo. It is easy to find because it is located immediately south of the bright Fomalhaut in Piscis Austrinus.

Its two brightest stars, Alpha (α) and Beta (β), contrast nicely, since Alpha is brilliant white, while Beta is noticeably orange—it is a red giant. Delta (δ) is a double star that can be separated by the naked eye; it is made up of yellow and red giant stars. For observers with larger telescopes, the region north of Theta (θ) has rich pickings for it is home to a small cluster of galaxies of the tenth and eleventh magnitudes.

LACERTA, THE LIZARD

Lacerta is a small northern constellation made up of a zigzag of faint stars. Its brighter head stars lie in the Milky Way, which, as always, is worth a sweep with binoculars. Quite a bright cluster shows up just to the west of Alpha (α).

PISCIS AUSTRINUS, THE SOUTHERN FISH

Gulping water cascading from the urn of Aquarius, this Southern Fish was considered to be the parent of the fishes that formed the zodiacal constellation Pisces. It is a small constellation, lifted from obscurity by its first-magnitude lead star Fomalhaut, meaning "fish's mouth" in Arabic. Because it appears in a region of generally faint stars, Fomalhaut stands out strikingly. The only other star worth looking at is Beta (β), which is a wide double.

◄ Little blobs of gas with tails form at the edge of the Helix
Nebula in Aquarius. Astronomers call them cometary knots.

NORTHERN HEMISPHERE

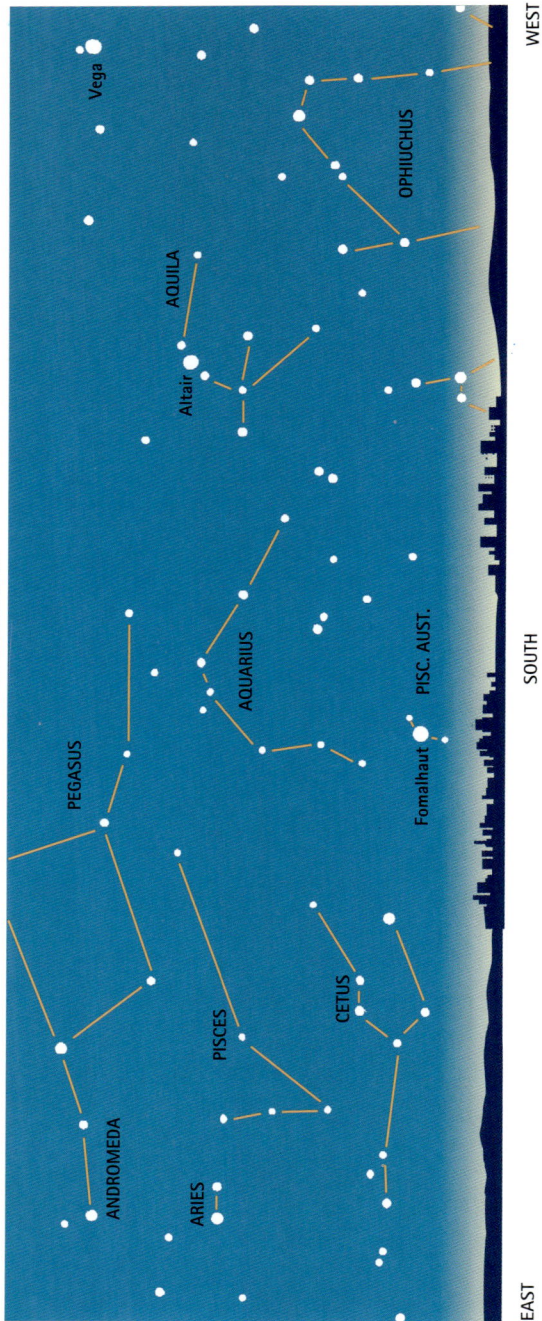

EAST

SOUTH

WEST

SEPTEMBER SKIES—LOOKING SOUTH
Constellations visible in North America and Europe at about 11:00 P.M. on about September 7.

The flying horse Pegasus continues winging its way west across the heavens, chasing the stars of the Summer Triangle. With Aquarius in mid-skies on the meridian, Pisces swimming behind, and Cetus rising, the heavens have taken on a watery aspect.

This is reinforced by the rising almost due south of Fomalhaut and the Southern Fish. Except for Fomalhaut and the three stars of the Summer Triangle, the southern skies look relatively barren.

NORTHERN HEMISPHERE

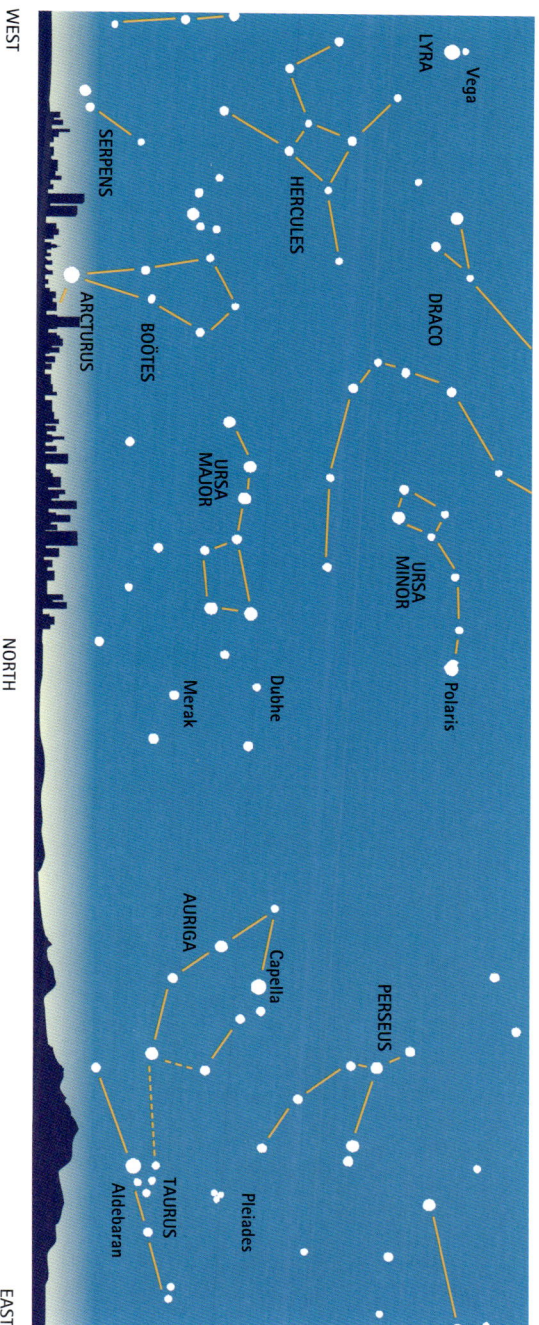

WEST

NORTH

EAST

SEPTEMBER SKIES—LOOKING NORTH

Constellations visible in North America and Europe at about 11:00 P.M. on about September 7.

This month the Big Dipper or Plow reaches its lowest point, with the handle nearly horizontal. The two stars that form the pointers to Polaris, Merak (lower) and Dubhe, are close to the meridian and pointing almost vertically upward. To the west,

Arcturus is close to setting and may be difficult to spot. In the east, however, Capella has been joined by the red eye of the bull, Aldebaran, as Taurus rises over the horizon. The Pleiades, located almost vertically above Aldebaran, are unmistakable.

Labels on chart: Vega, LYRA, SERPENS, HERCULES, DRACO, ARCTURUS, BOÖTES, URSA MAJOR, URSA MINOR, Polaris, Merak, Dubhe, AURIGA, Capella, PERSEUS, Pleiades, TAURUS, Aldebaran

SOUTHERN HEMISPHERE

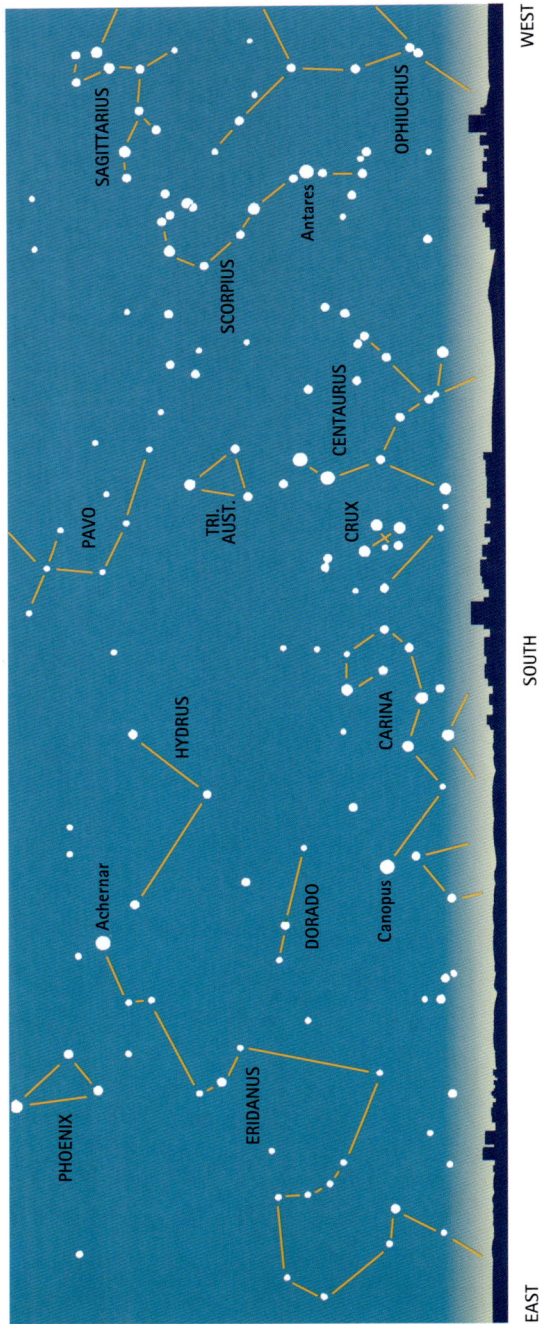

EAST

SOUTH

WEST

SEPTEMBER SKIES—LOOKING SOUTH

Constellations visible in Australia and South Africa at about 11:00 P.M. on about September 7.

Southern skies brighten this month as the Milky Way and its attendant brilliant constellations expand eastward. Canopus has reappeared in the southeast and is climbing. Its constellation—Carina—has drawn clear of the horizon. Centaurus, Scorpius, and Sagittarius all dazzle in the west. The east looks boring by comparison, occupied mainly by the widely meandering river Eridanus, whose "mouth" star, Achernar, now lies high above Canopus. The southern birds—phoenix, toucan, crane, and peacock—are now flying high.

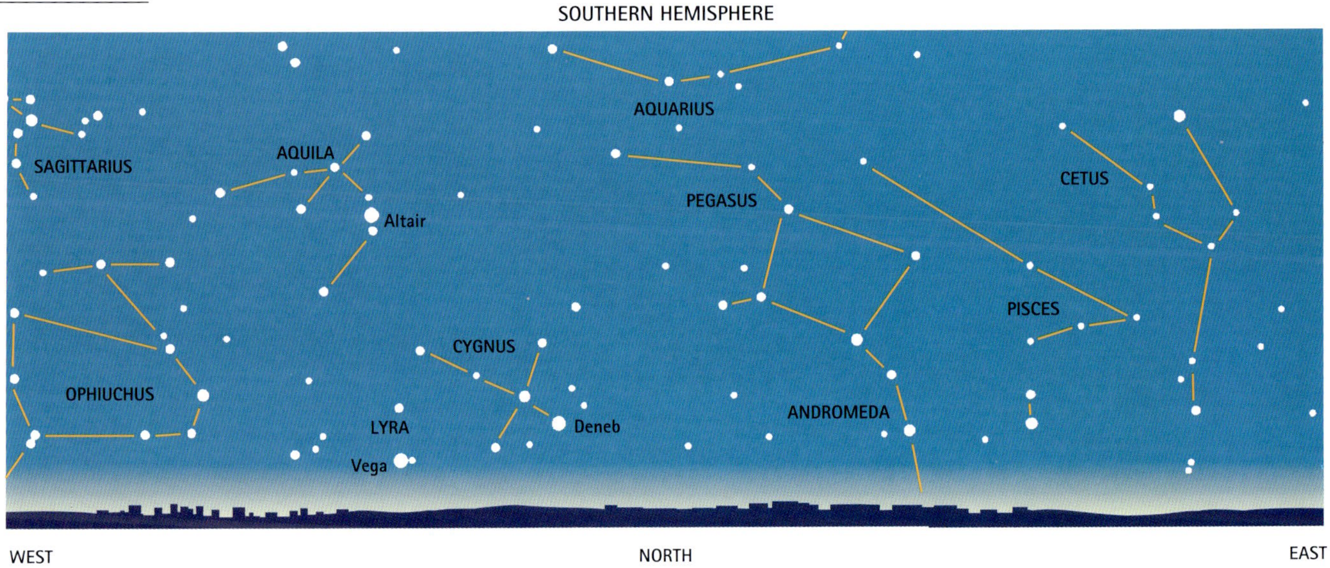

SOUTHERN HEMISPHERE

WEST

NORTH

EAST

SEPTEMBER SKIES—LOOKING NORTH

Constellations visible in Australia and South Africa at about 11:00 P.M. on about September 7.

The flying horse Pegasus continues winging its way west across the heavens, chasing the stars of the Summer Triangle. With Aquarius in mid-skies on the meridian, Pisces swimming behind, and Cetus risen, the heavens have taken on a watery aspect. This is reinforced by the appearance overhead (out

of the frame) of Fomalhaut and the Southern Fish. The northeast skies remain on the dull side; most interest is in the northwest, with the two birds, the eagle and the swan, conspicuous and nicely placed for observation.

October Stars

This month the Square of Pegasus sits right on the meridian and reminds us that fall is well advanced in the Northern Hemisphere and spring in the Southern. Few other constellations immediately strike the eye, as their stars are generally faint. Only in the far south does Fomalhaut continue to shine like a beacon.

PEGASUS, THE FLYING HORSE

One of the most ancient of the constellations, Pegasus has always been associated with the winged horse. The winged horse was a favorite theme among the artists of Assyria, one of the earliest civilizations in the Middle East. In Greek myths, Pegasus was the winged horse that leaped out of the corpse of the monstrous Medusa, after she had been beheaded by the hero Perseus.

The Square of Pegasus, the almost perfect square made by the stars Beta (β), Alpha (α), Gamma (γ), and Alpha in Andromeda, is one of the most recognizable star patterns in the heavens. All four stars are of about the same (second) magnitude. Of equal brightness is Epsilon (ε), just out of the frame here, but shown on the previous map (page 93). Named Enif, it represents the horse's nose. Enif is a wide double, visible

PEGASUS AND EQUULEUS.

▲ A fine portrayal of Pegasus, which could fly across the heavens.

OCTOBER SKIES ▶
Constellations visible near the meridian at about 11:00 P.M. during the first week in October.

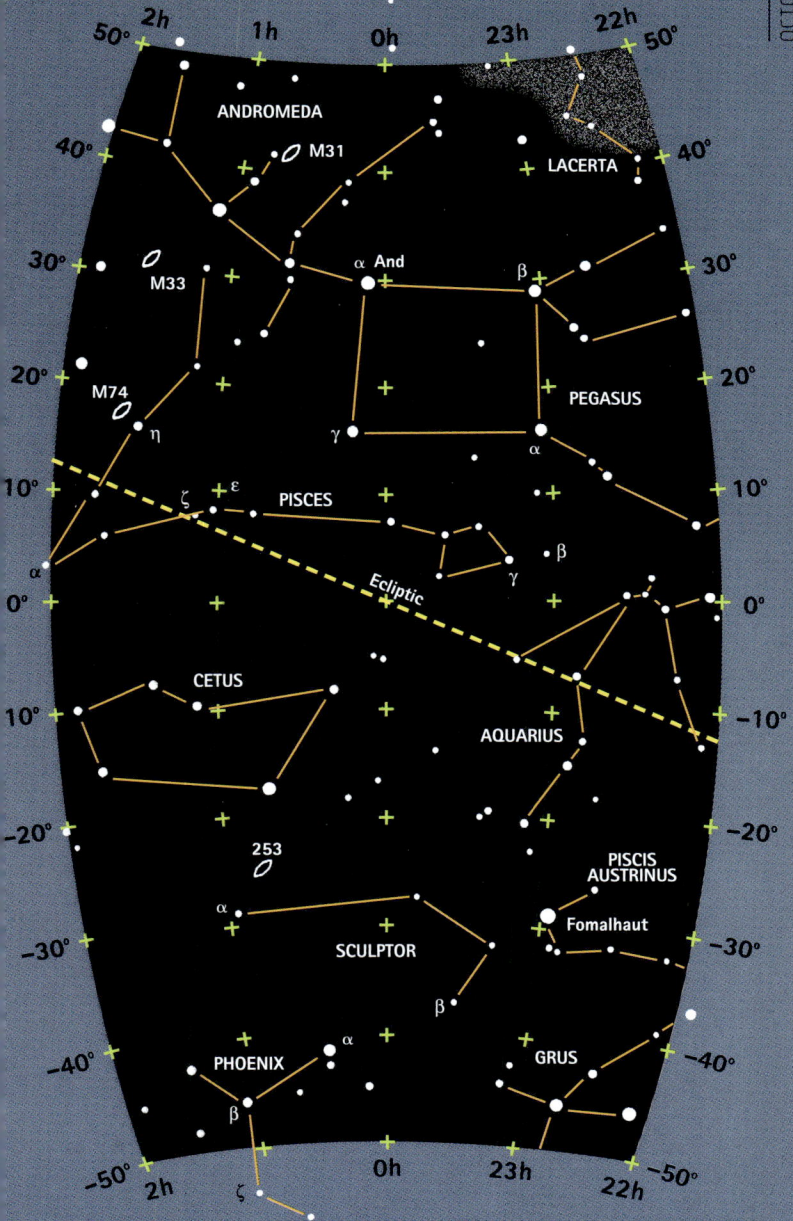

2h 1h 0h 23h 22h

50° 50°

ANDROMEDA

M31

40° 40°

LACERTA

30° M33 α And β 30°

PEGASUS

20° 20°

M74

η

10° ζ ε PISCES 10°

Ecliptic

γ α β

0° 0°

α γ

CETUS

10° AQUARIUS −10°

PISCIS
AUSTRINUS

−20° 253 −20°

α Fomalhaut

SCULPTOR

−30° β −30°

PHOENIX α GRUS

−40° β −40°

−50° 2h ζ 0h 23h 22h −50°

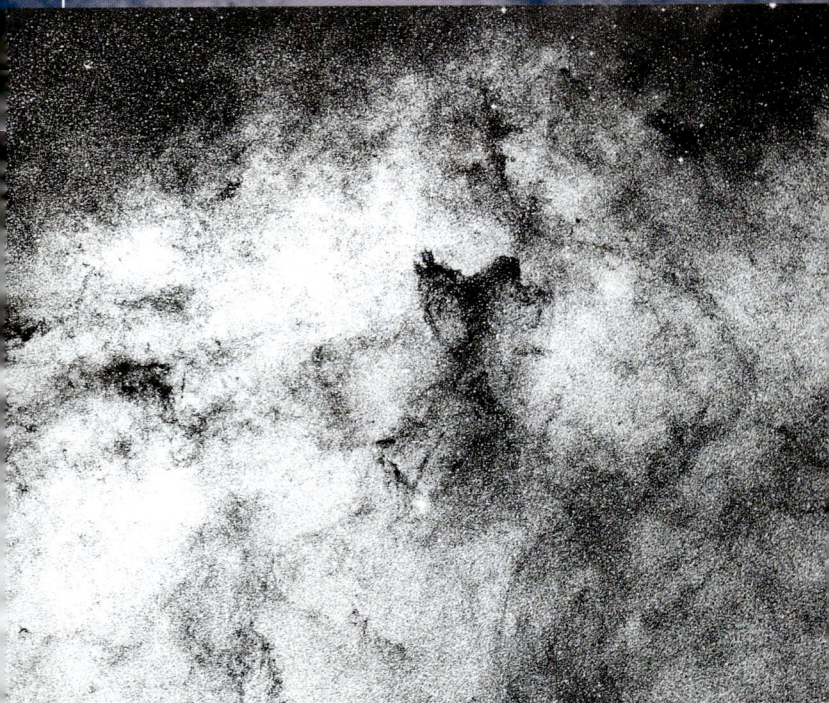

▲ In far-north October skies, the Milky Way is brilliant.

in binoculars. A short distance away from Enif is the fine globular cluster M15. Visible in binoculars, it resolves into individual stars in larger telescopes. It has a highly luminous core, which strongly emits X-rays. This leads astronomers to believe that there must be a black hole lurking within it.

PHOENIX, THE PHOENIX

This southern bird lies close to another, Grus, the crane. A relatively modern constellation (1500s), it is named after the fabled bird that was reincarnated from its own ashes. After living for 500 years, it burned itself on a funeral pyre, but flew out of the flames reborn. It has one or two interesting stars, including Beta (β) and Zeta (ζ), which are both doubles. Zeta is the easier to separate. Its brighter component varies in brightness every 40 hours, since it is an eclipsing binary.

PISCES, THE FISHES

This is a large, but faint constellation of the zodiac. The Sun passes through it between March 12 and April 18 every year. In mythology, the two fishes represent Venus and her son Cupid. They had to dive into the River Euphrates to escape from the clutches of the monster Typhon.

Pisces is the constellation in which the ecliptic and celestial equator intersect, or, in other words, where the Sun appears to cross the celestial equator as it travels north. This happens on about March 21 every year. This is the time of the spring, or vernal equinox, which marks the beginning of spring in the Northern Hemisphere and of the fall in the Southern.

The intersection of the ecliptic and the celestial equator is also the starting point (0 hours) for celestial longitude, or right ascension (see page 8).

All that having been said, the constellation from an observer's viewpoint is of little worth. For a start, the V-shaped lines of stars that make up the constellation are difficult to make out. In small telescopes Alpha (α) is a binary star, while Zeta (ζ) is a wide double. Among the several galaxies found in Pisces, M74, immediately east of Eta (ε), is within reach of small telescopes, visible as a misty patch.

SCULPTOR, THE SCULPTOR

Sculptor is yet another faint constellation in this part of the sky, found by scanning east from the bright Fomalhaut in the neighboring Piscis Austrinus. This is a modern constellation, introduced in the 1700s, and has no mythological connections. Its main attractions are its galaxies, most of which require larger telescopes to make them out. An exception is NGC253 to the north of Alpha (α). At the seventh magnitude, this can be seen easily with binoculars. In telescopes it looks cigar-shaped because we are looking at it sideways.

One of the Sculptor ▶ group of galaxies. This is NGC253, a spiral galaxy that we see sideways. Our own galaxy would look much like this from a distance.

NORTHERN HEMISPHERE

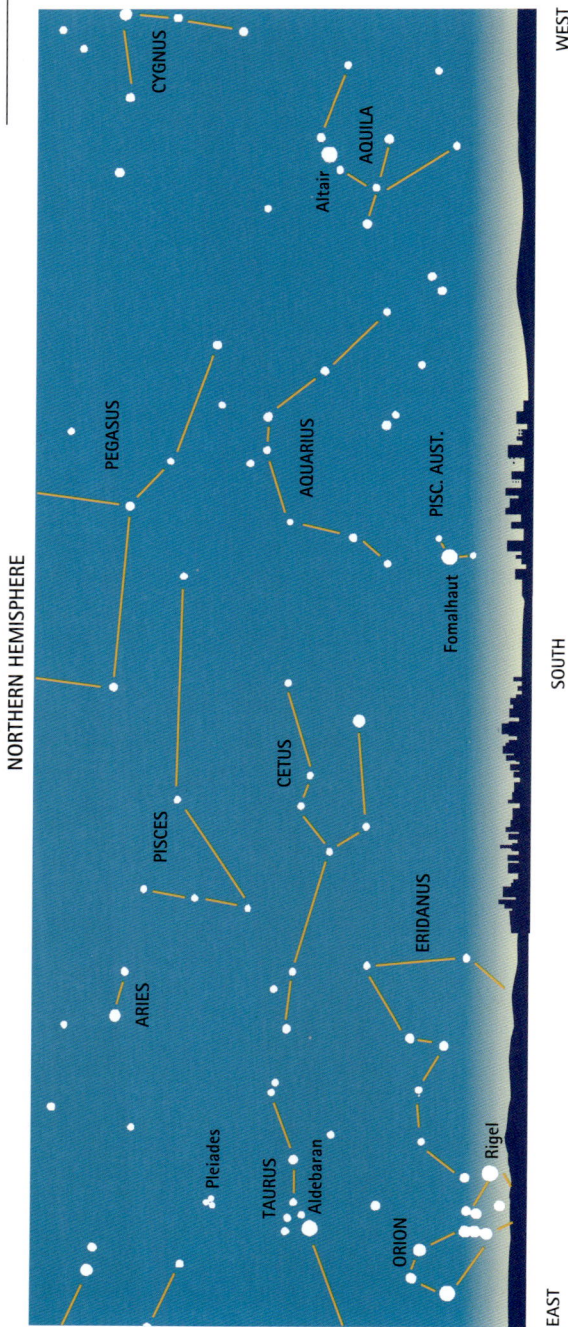

EAST

WEST

SOUTH

CYGNUS
AQUILA
Altair
PEGASUS
AQUARIUS
PISC. AUST.
Fomalhaut
PISCES
CETUS
ARIES
ERIDANUS
Pleiades
TAURUS
Aldebaran
ORION
Rigel

OCTOBER SKIES—LOOKING SOUTH
Constellations visible in North America and Europe at about 11:00 P.M. on about October 7.

This month the Square of Pegasus sits on the meridian and reminds us that the fall is upon us. Generally, though, the skies are far from brilliant, occupied by Pegasus, Andromeda (just out of the frame), Pisces, Aries, Aquarius, Cetus, and the meandering river Eridanus, which has appeared over the southeast horizon. But, as the weather cools and skies get darker, we see Taurus risen and Orion peeping over the horizon, hinting at more brilliant times to come.

NORTHERN HEMISPHERE

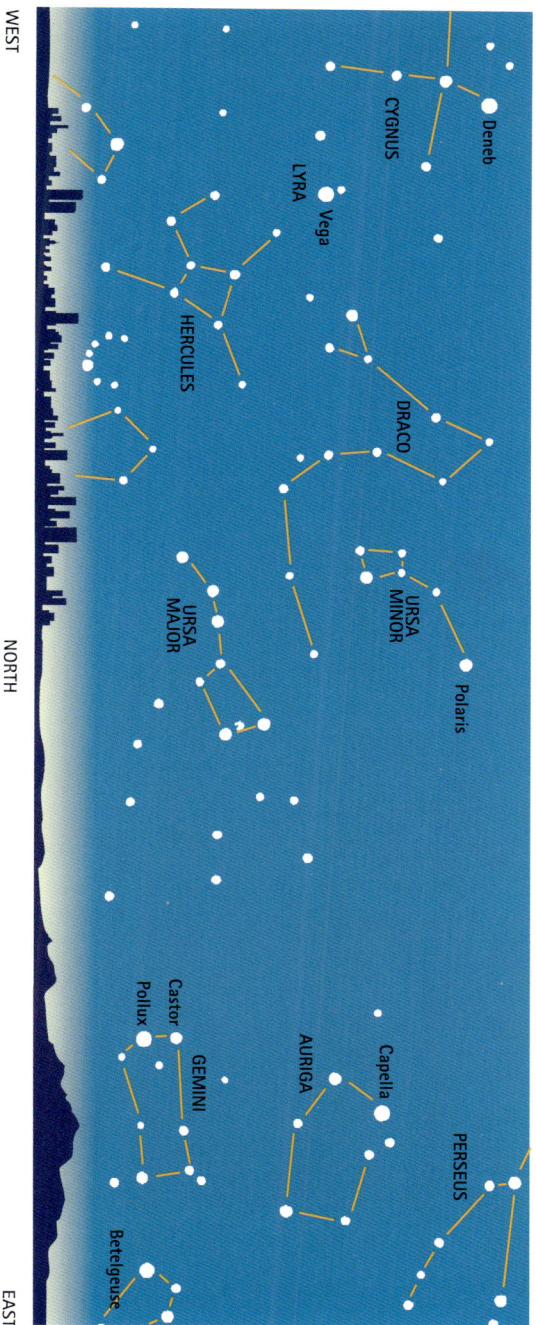

WEST

NORTH

EAST

OCTOBER SKIES—LOOKING NORTH

Constellations visible in North America and Europe at about 11:00 P.M. on about October 7.

In the west, the Summer Triangle of Deneb, Vega, and Altair is making its descent, telling us that summer is long gone. In the east, Gemini has risen, with its two lead stars Castor (top) and Pollux one above the other. At the same height as Pollux and further east, the noticeably red Betelgeuse puts in an appearance as Orion begins to rise. Northern observers can once again look forward to sampling the ample delights of one of their favorite constellations.

SOUTHERN HEMISPHERE

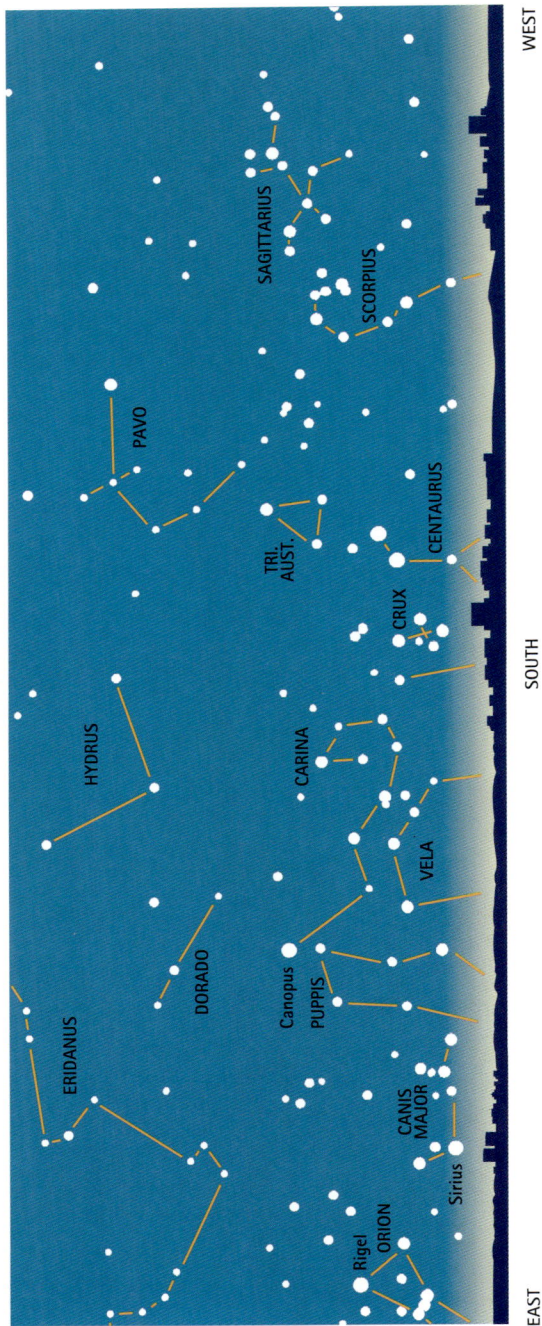

EAST

SOUTH

WEST

OCTOBER SKIES—LOOKING SOUTH

Constellations visible in Australia and South Africa at about 11:00 P.M. on about October 7.

From the horizon up to mid-skies, the heavens this month are outstanding. Orion is rising in the east, with Rigel well above the horizon. Canis Major has mostly risen too, with Sirius climbing into view. Canopus leads the march of the three nautical constellations—keel (Carina), poop (Puppis), and sails (Vela). Crux is close to the meridian low down. Antares and the head of Scorpius have disappeared over the horizon, but the curve of stars that outline the scorpion's deadly stinging tail are still in view.

SOUTHERN HEMISPHERE

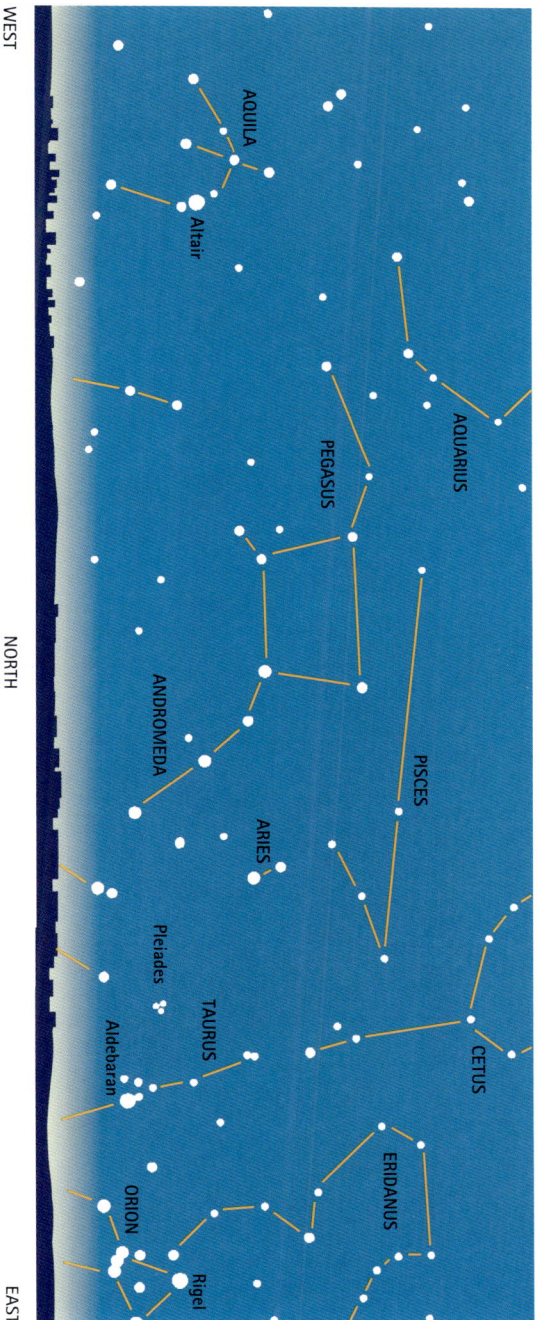

WEST   NORTH   EAST

OCTOBER SKIES—LOOKING NORTH

Constellations visible in Australia and South Africa at about 11:00 P.M. on about October 7.

This month the Square of Pegasus sits on the meridian and reminds us that spring is here. The northern aspects of the skies, however, are far from brilliant, with most of the heavens occupied by relatively faint constellations—Pegasus, Andromeda,

Pisces, Aries, Aquarius, Cetus, and the meandering river Eridanus. But we know that there are more brilliant times ahead, with Taurus and Orion rising in the east. This month is a good time to observe the Andromeda Galaxy, close to the meridian.

November Stars

Meridian skies are far from spectacular this month. Pegasus has slipped away west, and Taurus and Orion are still waiting in the wings in the east. Andromeda, however, provides a spectacular highlight, revealing as a faint misty patch the galaxy that represents the furthest object in the heavens the naked eye can see, at a distance of 12,000,000,000,000,000,000 miles (20,000,000,000,000,000,000 km).

ANDROMEDA

While the pattern of stars making up this constellation is not particularly memorable, Andromeda can easily be found because it is joined to one of the most distinctive star groups in the sky—the Square of Pegasus. But in mythology the two figures represented by the constellations were not linked. Andromeda was the lovely daughter of King Cepheus and Queen Cassiopeia. To cut a long mythological story short, the sea god Poseidon became upset when Cassiopeia boasted about how

▲ The central region of the Andromeda Galaxy, one of our galactic neighbors.

NOVEMBER SKIES ►
Constellations visible near the meridian at about 11:00 P.M. during the first week in November.

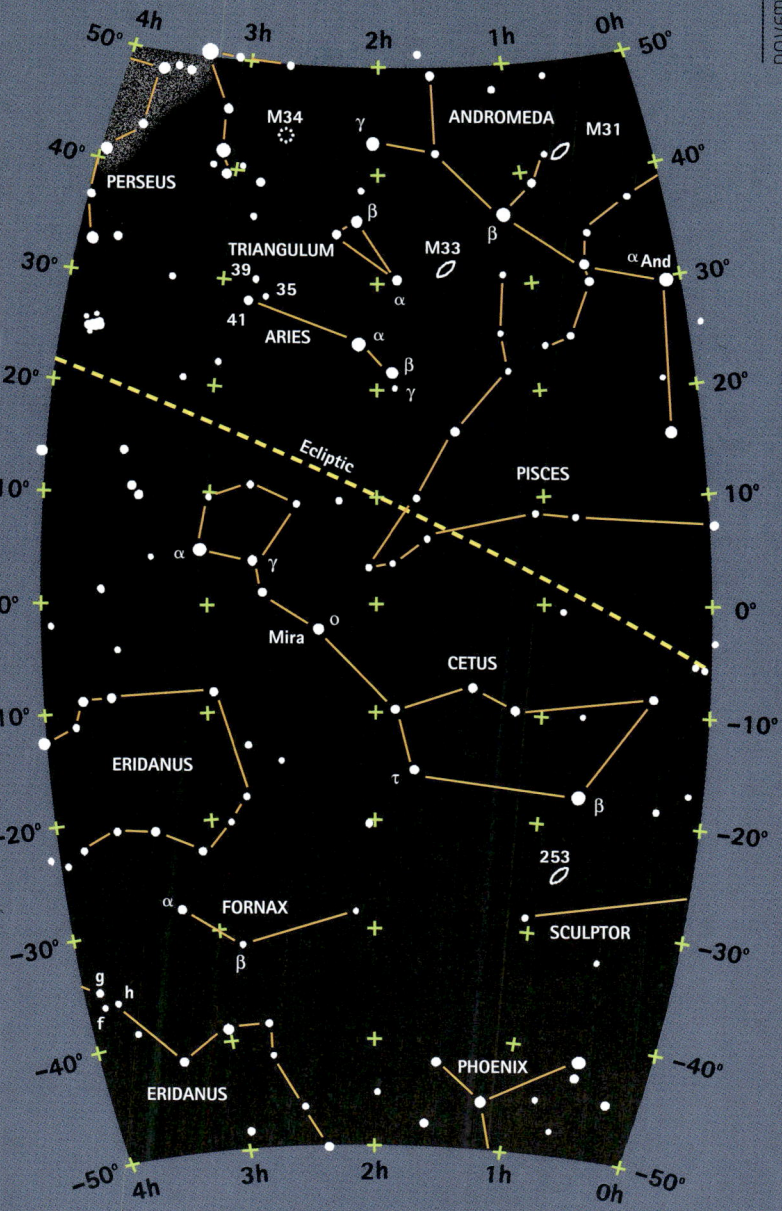

50° 4h   3h   2h   1h   0h 50°

40°

M34   γ   ANDROMEDA   M31

PERSEUS

TRIANGULUM   β
39   M33   α And
35   α
41
ARIES   α
β
γ

Ecliptic   PISCES

α   γ

o
Mira

CETUS

ERIDANUS

τ
β

253

α   FORNAX

β   SCULPTOR

g
h
f
ERIDANUS   PHOENIX

-50° 4h   3h   2h   1h   0h -50°

beautiful she was. As a sacrifice to placate the god, poor Andromeda was chained to a rock to provide a meal for the sea monster Cetus. All ended happily, however, when the hero Perseus happened by and snatched the fair maiden from the monster's jaws in the nick of time.

Perhaps the most appealing of the constellation's stars is Gamma (γ), which is an easy double for small telescopes. The two components make a lovely pair, the one bright orange, the other blue-green. But it is the fuzzy patch seen in the north of Andromeda that merits the most attention. Once thought to be a gas cloud in our own galaxy, it was named the Great Nebula in Andromeda. But the increasingly powerful telescopes that came into use early this century showed that this fuzzy patch, which Messier cataloged as M31, is another galaxy. It lies far beyond the stars of our own galaxy at a distance of nearly 2.5 million light-years.

The Andromeda galaxy is one of the few that can be seen with the naked eye. Binoculars or a small telescope show it better, of course, but a large telescope is needed to bring out the galaxy's spiral arms. The smaller instruments, however, will reveal its two close companion galaxies, M32 and NGC295. The Andromeda galaxy is one and a half times as big as our own galaxy and indeed is the biggest galaxy in the cluster we call the Local Group.

ARIES, THE RAM

Aries is one of the smaller constellations of the zodiac, sandwiched between Taurus and Pisces. The Sun travels through Aries between April 18 and May 14 every year. Aries represents the ram with the Golden Fleece, sought by Jason and the Argonauts.

It is not a conspicuous constellation and is best located by reference to the Square of Pegasus in the west and the Pleiades in the east. Gamma (γ) is an easy double for small telescopes. Historically, it is interesting as one of the first doubles to be detected, by the English scientist Robert Hooke in 1664. The three stars 35, 39, and 41 form a little triangle that was once known as Musca Borealis, or the Northern Fly, representing flies buzzing around the Ram's tail.

CETUS, THE WHALE

Also known as the Sea Monster, this is one of the largest constellations. It represents the monster that was about to devour the fair Andromeda (see opposite).

Cetus is not easy to trace because its stars are mostly faint; even its brightest—Alpha (α) and Beta (β)—are only third magnitude. Gamma (γ) is a lovely binary for small telescopes, with blue and yellow components. Tau (τ) is of interest because it is a yellow dwarf star that is nearly

A star–forming region in the beautiful spiral galaxy M33. ▲

identical to the Sun. At a distance of 11.9 light-years, it is one of the nearest Sunlike stars.

But the *pièce de résistance* of the constellation is the star Omicron (o), better known as Mira, meaning "the Wonderful." It got its name because it was the first red-giant variable star to be discovered, by the Dutch astronomer David Fabricus in 1596. Mira varies in brightness noticeably over a period of about 11 months. At maximum, it is easily visible to the naked eye at around the third magnitude, but it fades at minimum to about the tenth magnitude, when it becomes difficult to spot even with binoculars.

TRIANGULUM, THE TRIANGLE

This small constellation could not be better named, as its trio of main stars form a perfect right-angled triangle. The object to look for in Triangulum is M33, a spiral galaxy that presents itself to us head-on. It is just bright enough to spot with the naked eye on a really dark night, between Alpha (α) and the star Beta (β) in Andromeda.

NORTHERN HEMISPHERE

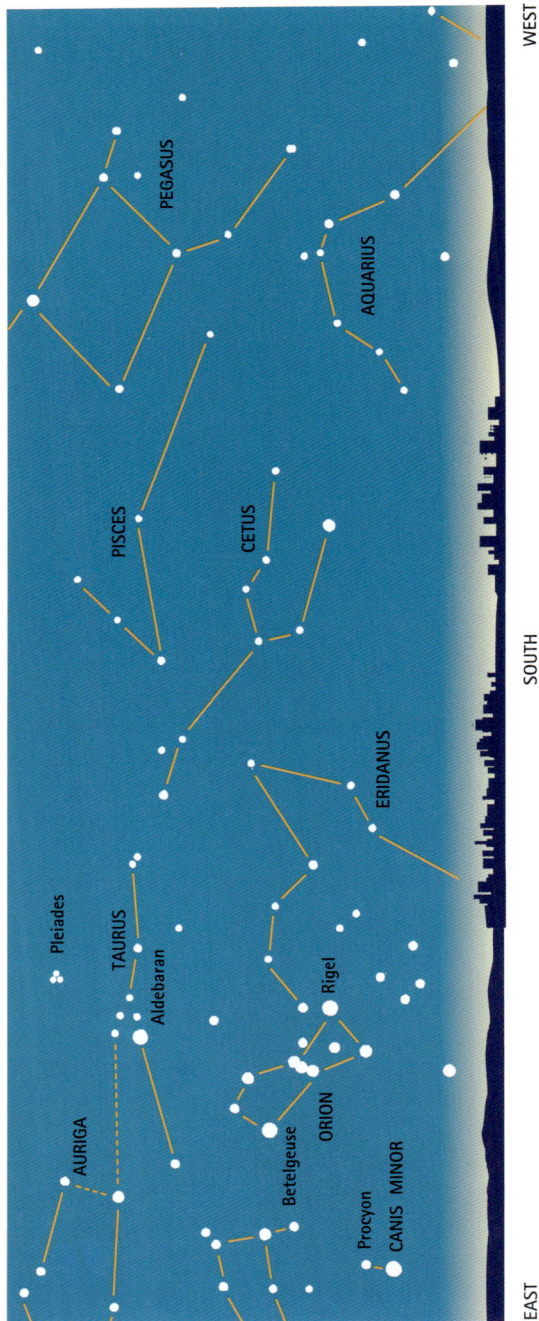

EAST   SOUTH   WEST

NOVEMBER SKIES—LOOKING SOUTH
Constellations visible in North America and Europe at about 11:00 P.M. on about November 7.

While Pegasus is flying ever westward—along with the dull constellations of Pisces, Cetus, and Aquarius—magnificent Orion is striding across the heavens in the east. He faces the charging bull Taurus, with horns lowered ready to gore, and

glaring eye, marked by Aldebaran. Slightly higher in the sky are the Pleiades, the star group named after the sisters whom Orion chased in Greek mythology. This outstanding open, or galactic cluster, is now well placed for observation.

NORTHERN HEMISPHERE

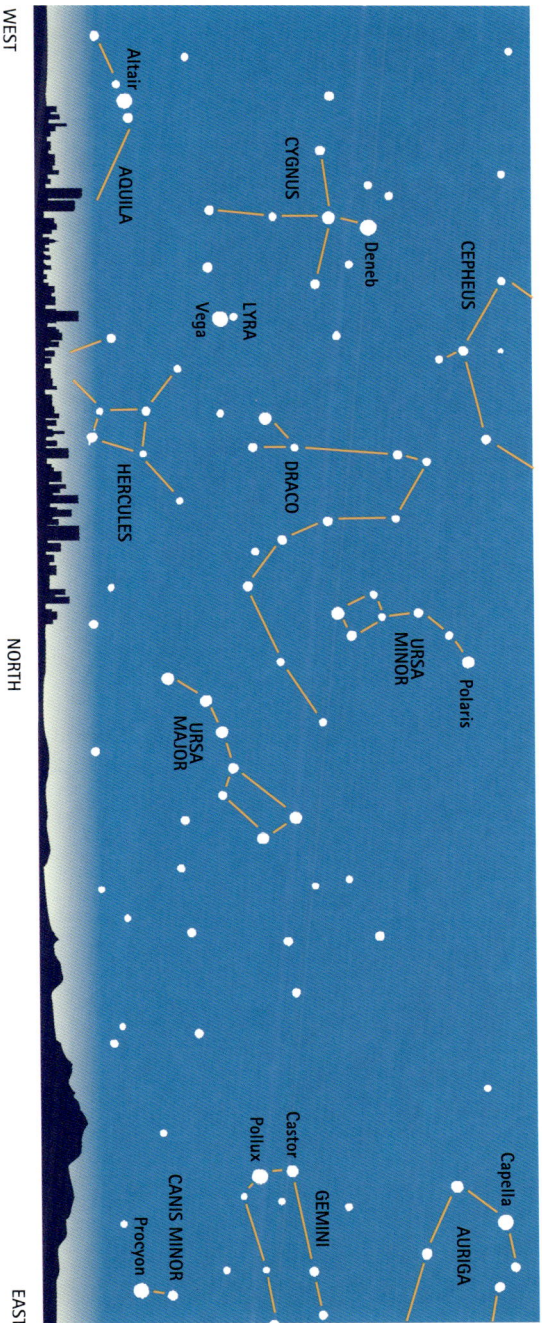

WEST

NORTH

EAST

NOVEMBER SKIES—LOOKING NORTH

Constellations visible in North America and Europe at about 11:00 P.M. on about November 7.

This month the swan (Cygnus) is flying vertically in the west, with its long neck outstretched and wings spread wide, as if diving headlong for the horizon. However, the other bird, the eagle (Aquila), will get there first. Their bright stars

still form with Vega (Lyra) a conspicuous triangle that a few months ago was high overhead in the balmy night skies of summer. In the east, Canis Minor has risen, and Procyon joins Castor, Pollux, and Capella.

SOUTHERN HEMISPHERE

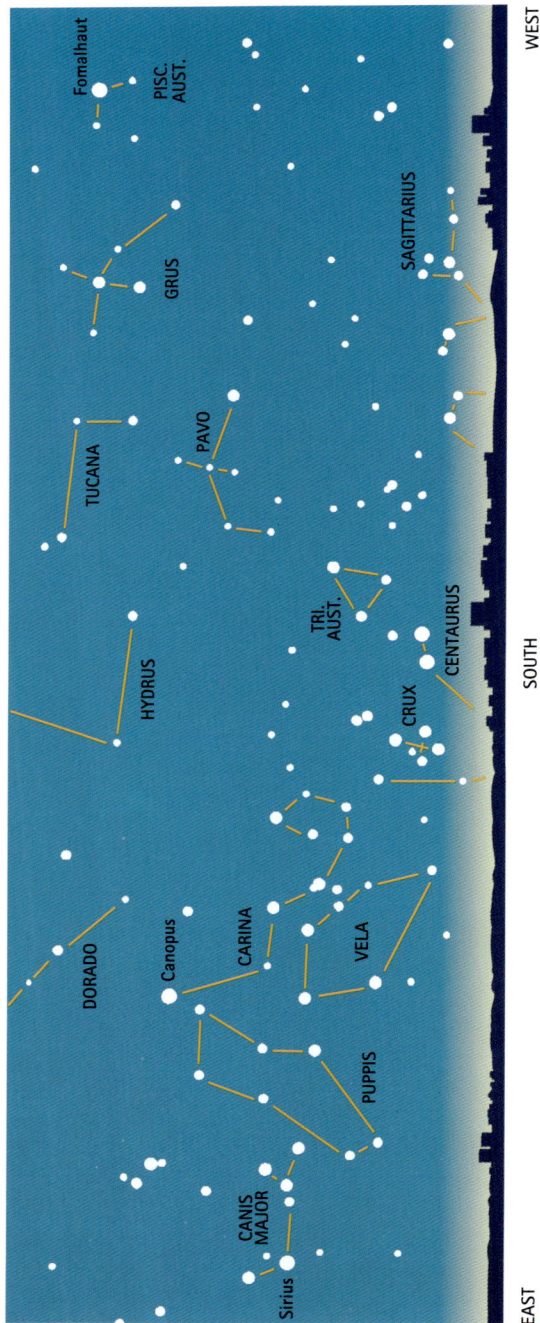

EAST

SOUTH

WEST

Fomalhaut
PISC. AUST.
GRUS
TUCANA
PAVO
SAGITTARIUS
HYDRUS
TRI. AUST.
CENTAURUS
CRUX
DORADO
Canopus
CARINA
VELA
PUPPIS
CANIS MAJOR
Sirius

NOVEMBER SKIES—LOOKING SOUTH
Constellations visible in Australia and South Africa at about 11:00 P.M. on about November 7.

The lower part of the skies are not quite as dazzling this month. Scorpius and Sagittarius have mostly disappeared in the southwest, taking with them the most brilliant region of the Milky Way, for it is in the direction of Sagittarius that the center of our galaxy is located. The southeast is now the brightest aspect of the skies, with Sirius and Canopus relatively close. Low in the south the twin bright stars Alpha and Beta Centauri lie close to the meridian, while Crux has just passed it.

SOUTHERN HEMISPHERE

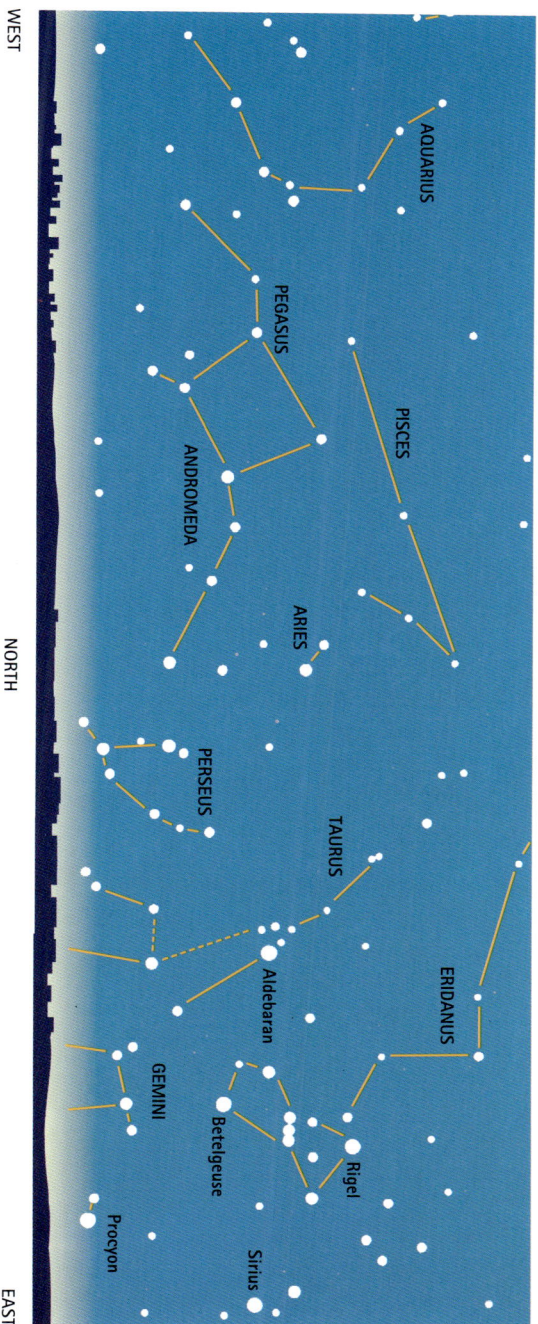

WEST

NORTH

EAST

NOVEMBER SKIES—LOOKING NORTH

Constellations visible in Australia and South Africa at about 11:00 P.M. on about November 7.

While Pegasus is flying ever westward—along with the dull constellations of Pisces, Aries, and Aquarius—magnificent Orion is striding across the heavens in the east. He faces the charging bull Taurus, with horns lowered ready to gore, and glaring eye, marked by Aldebaran. Risen in the east now are the two celestial dogs Canis Major and Minor (just out of frame, far right). Brightest-star-in-the-sky Sirius is well above the horizon, although Procyon is low down. The Andromeda Galaxy is again well-placed for observation well clear of the horizon.

December Stars

This month the faint stars of the past few months have all but drifted away. And dazzling constellations and the Milky Way are moving in to take their place, from Perseus in the north, through Taurus and Orion, to Canis Major in the south. The night sky's most famous cluster, the Seven Sisters, or the Pleiades, is now well placed for observation from both hemispheres.

ERIDANUS

This immensely long constellation winds itself through the southern heavens like a river meandering its way from its source to the sea. It is no small wonder, therefore, that Eridanus has been identified with rivers from the earliest times—by the Babylonians and Egyptians as the Euphrates and the Nile in the Middle East, and by the Romans as the Po in Italy.

Eridanus rises near the bright Rigel in Orion, and leisurely, with much winding, makes its way south. Its mouth is marked by its brightest star, the first-magnitude Achernar, meaning "end of the river" in Arabic. This star is the ninth brightest in the sky.

The winding constellation has a few interesting stars. Omicron-2 (σ^2) is a multiple star, which small telescopes will readily resolve into a pair. In larger telescopes, the fainter of the two reveals itself to be a pair of dwarfs, one red, the other white. Further west lies Epsilon (ϵ), which is the third-nearest naked-eye star to us, at a distance of 10.7 light-years. It is also the nearest Sunlike star.

FORNAX, THE FURNACE

A modern constellation (1750s), the furnace commemorates the advanced iron-smelting furnace that was helping to drive the Industrial Revolution at the time. It is not a bright constellation, and is of interest mainly to telescopic observers. They can spot southwest of Beta (β) a small irregular galaxy called the Fornax system, which is a member of our Local Group. Southeast of Beta, near the triangle of stars f, g, and h in neighboring Eridanus, is the Fornax cluster of galaxies. At a distance of some 55 million light-years, it is one of the nearest clusters to us.

DECEMBER SKIES ►
Constellations visible near the meridian at about
11:00 P.M. during the first week in December.

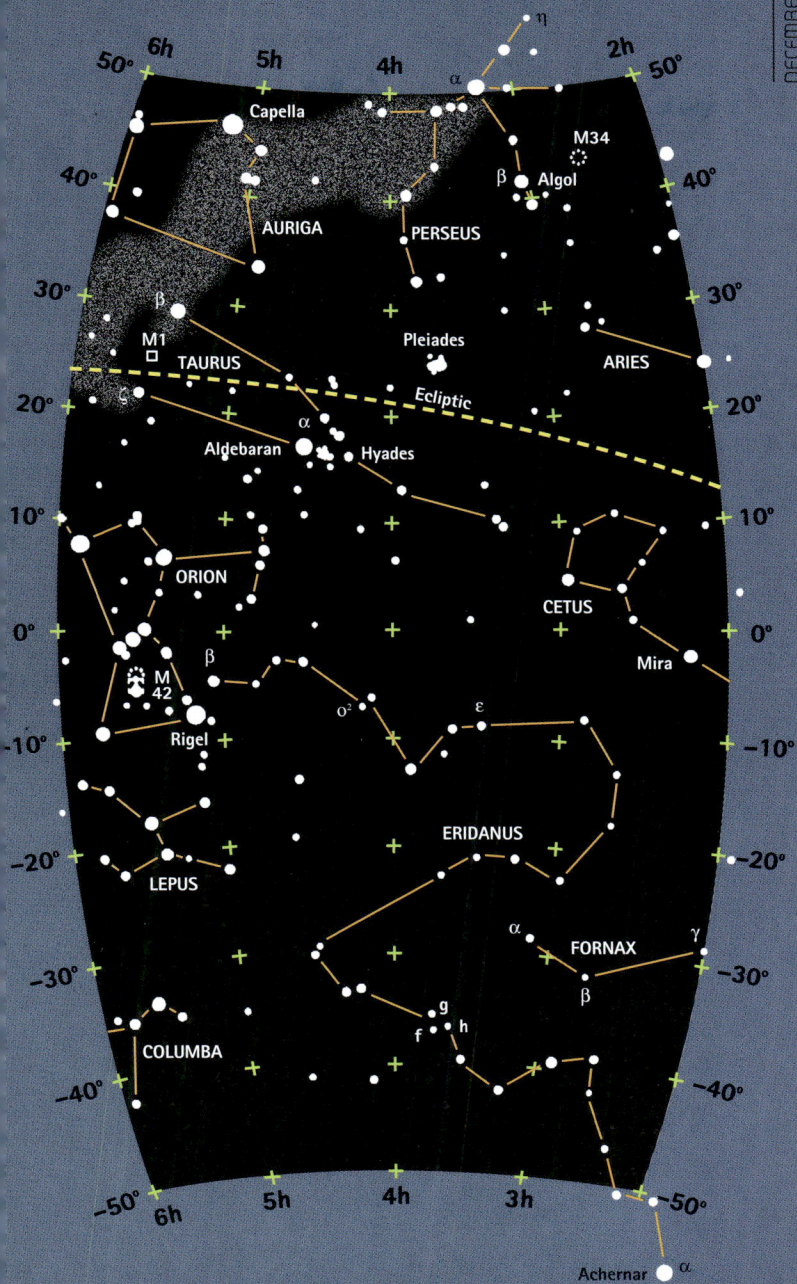

▲ The Double Cluster looks delightful in binoculars.

PERSEUS

Perseus was one of the great heroes of Greek mythology, along with the likes of Hercules and Orion. It was he who beheaded the dreaded Medusa, the Gorgon who had serpents for hair and whose gaze would turn mortals into stone. And it was he who, heading home after this adventure, rescued Andromeda from the jaws of the sea monster Cetus.

This far northern constellation has much to commend it. The asterism formed by Beta (β) and its neighbors represents the head of Medusa. Beta is famed as the "Winking Demon" Algol. It is a variable star of the eclipsing binary type, in which a bright and a dim star orbit each other in our line of sight and periodically eclipse each other. Beta's brightness therefore, dips periodically, making it appear to "wink." This happens as regularly as clockwork every 2 days, 21 hours, when it dims from the second to the third magnitude for about 19 hours. Algol was the first eclipsing binary to be recognized, by the English astronomer John Goodricke in 1782. Northwest of Algol and about halfway to Gamma (γ) in neighboring Andromeda is an open cluster (M34) that is just visible to the naked eye.

The body of Perseus lies in the Milky Way, and there are many delights to be seen there. Alpha (α) is the brightest of a loose association of stars known as Melotte 20, and they look impressive with binoculars. At the most northerly extremity of the constellation, beyond Eta (η), is the famous Double Cluster, a pair of open clusters just visible to the naked eye and beautiful through binoculars or small telescopes.

Nebulosity surrounds the stars in the Pleiades cluster. They are white, hot, and young. ▼

TAURUS, THE BULL

One of the easiest constellations of the zodiac to spot, Taurus can with only a little imagination be visualized as a charging bull, with baleful red eye and lowered horns. The Sun travels through the constellation between May 14 and June 21 every year.

In mythology, Taurus is the bull that features in one of the many seductions of Zeus. To capture the attention of the lovely Princess Europa, he turned himself into a handsome, white bull. She climbed on his back, and he swam to the island of Crete, where he resumed his manly form and loved her. One of their children became King Minos, who built the famous palace at Knossus and encouraged bull worship.

Highlights abound in Taurus. Aldebaran is the noticeably reddish eye of the Bull. It is surrounded, in sight but not in space, by the open star cluster called the Hyades. This cluster has a prominent V-shape and is easily seen with the naked eye, with its brightest stars of about the fourth magnitude. At a distance of some 130 light-years, the Hyades is one of the nearest open clusters (but is twice as far away as Aldebaran).

Farther north is the even more spectacular open cluster of the Pleiades (M45). It is commonly called the Seven Sisters, although exceptional eyesight is needed to make out its seven brightest stars. Alcyone (magnitude 2.9) is the brightest. In binoculars, many more stars show up in the cluster, which is estimated to contain as many as 300 stars in all. They are comparative youngsters, around 10 to 20 million years old. They are hot and white, and are surrounded by misty nebulosity, which photographs bring out (see picture on page 119). The center of the Pleiades cluster lies about 400 light-years away, much farther than the older Hyades.

Another highlight of the constellation, however, needs a telescope to make out, just north of Zeta (ζ), the star at the tip of the Bull's southern horn. It is the first object in Messier's Catalogue, M1, or the Crab Nebula. It is what remains of a supernova explosion that took place on July 4 AD1054 and was witnessed by Chinese astronomers. They recorded that this "guest star" shone so brightly that it could be seen in the daylight for more than three weeks. This nebulous supernova remnant is still expanding rapidly. Embedded within it is what remains of the star that went supernova—a tiny pulsar. The Crab pulsar was the first pulsar to be detected visually, and beams flashes of light in our direction 30 times every second.

The Crab Nebula in Taurus was ▶
born in a supernova, spotted
nearly 1,000 years ago.

NORTHERN HEMISPHERE

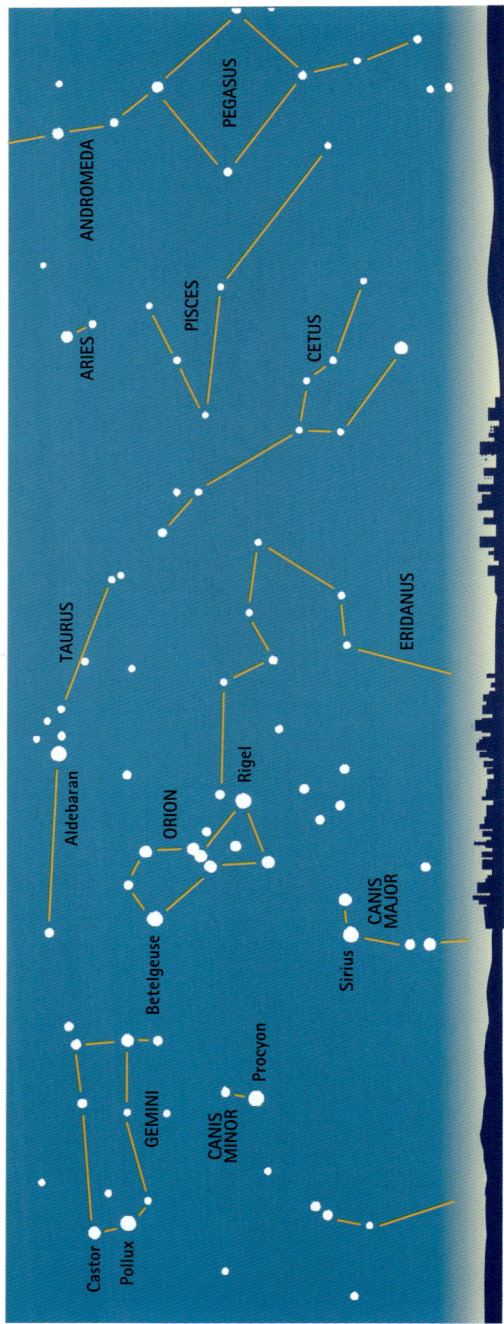

WEST

PEGASUS

ANDROMEDA

ARIES

PISCES

CETUS

TAURUS

Aldebaran

ORION

Rigel

ERIDANUS

SOUTH

Betelgeuse

CANIS
MAJOR

Sirius

GEMINI

CANIS
MINOR

Procyon

Castor
Pollux

EAST

DECEMBER SKIES—LOOKING SOUTH

Constellations visible in North America and Europe at about 11:00 P.M. on about December 7.

In the northern heavens this month, there is a two-way split. To the west of the meridian, the skies are relatively dull, dominated by Cetus and the winding Eridanus, as well as Pisces, Andromeda, and Pegasus. But what a different story it is to the east of the meridian. Orion is in mid-skies, with Taurus higher up and Canis Major lower down. Canis Minor and Gemini are also well above the horizon. Sirius, Rigel, Aldebaran, Capella, Pollux, and Procyon form a polygon of bright stars often called the winter hexagon.

NORTHERN HEMISPHERE

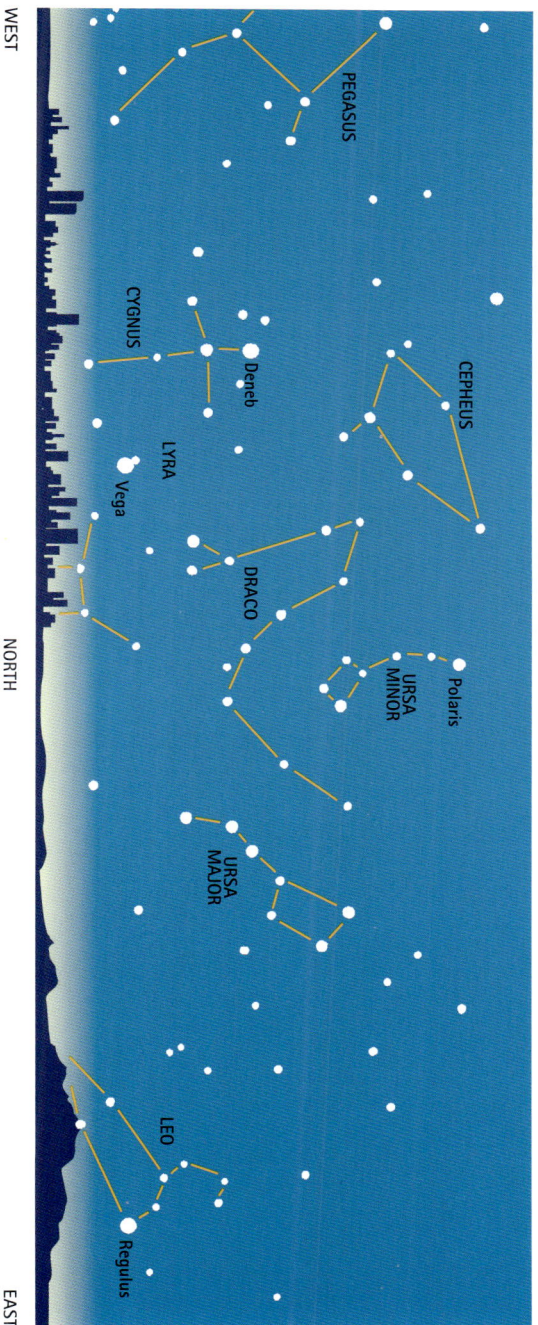

WEST

NORTH

EAST

DECEMBER SKIES—LOOKING NORTH
Constellations visible in North America and Europe at about 11:00 P.M. on about December 7.

Cygnus has not yet reached the western horizon, but Aquila has now disappeared beneath it. Vega is still visible low down further east, but the Summer Triangle is no more. After all, it is mid-winter. The handle stars of the Little Dipper are now on the

meridian, directly below Polaris. In the far west, the Square of Pegasus is prominent, but not for long. In the far east, Leo is rising, with bright Regulus and the curve of stars above it making the familiar sickle shape.

SOUTHERN HEMISPHERE

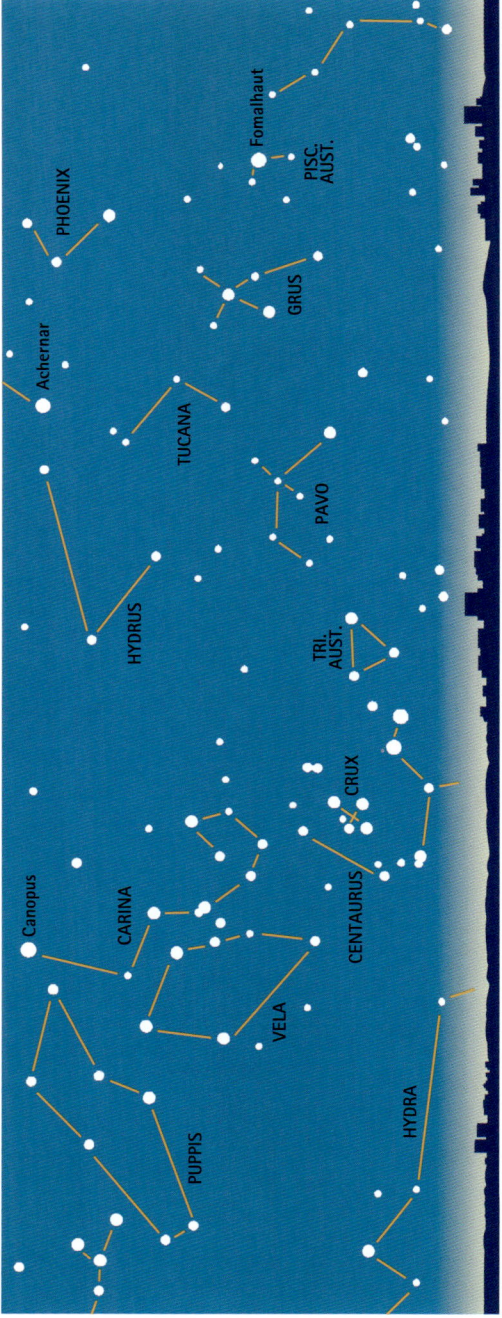

WEST

EAST

SOUTH

PISC. AUST.

Fomalhaut

PHOENIX

GRUS

Achernar

TUCANA

PAVO

HYDRUS

TRI. AUST.

CRUX

CENTAURUS

Canopus

CARINA

VELA

HYDRA

PUPPIS

DECEMBER SKIES—LOOKING SOUTH

Constellations visible in Australia and South Africa at about 11:00 P.M. on about December 7.

The southern triangle (Triangulum Australe) is back on the meridian this month, low down. Six months ago, in June, it was also on the meridian, but high up. This reminds us that the heavens have turned half-circle since then. Most of the interest this month lies in the southeast, where the brilliant constellations now lie. West of the meridian the skies are relatively dull, with only Achernar and Fomalhaut shining out among the flock of southern birds.

SOUTHERN HEMISPHERE

WEST   NORTH   EAST

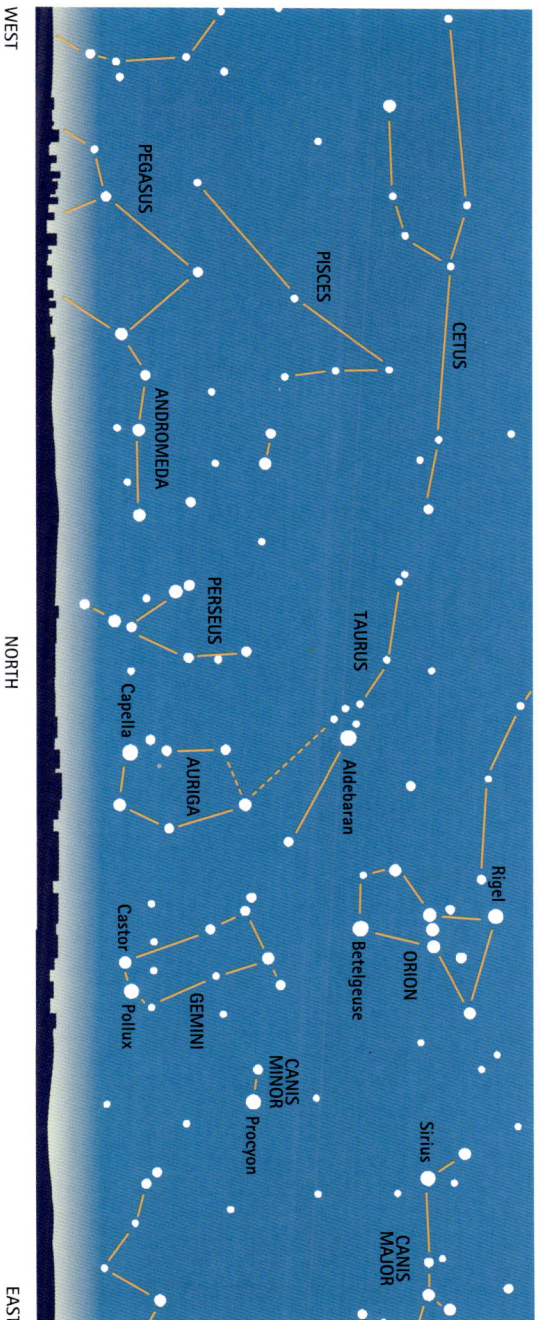

DECEMBER SKIES—LOOKING NORTH
Constellations visible in Australia and South Africa at about 11:00 P.M. on about December 7.

In the northern heavens this month, there is a two-way split. To the west of the meridian, the skies are relatively dull, dominated by sprawling Cetus and Pisces, as well as Andromeda and Pegasus. But what a different story it is to the east

of the meridian. Taurus is in mid-skies, with Orion higher still and Canis Major further east. Canis Minor, Gemini, and Auriga are also well above the horizon. Sirius, Rigel, Aldebaran, Capella, Pollux, and Procyon form a hexagon of bright stars.

Index